ADVANCES IN

CHROMATOGRAPHY

Volume 12

ADVANCES IN
CHROMATOGRAPHY

Volume 12

Edited by

J. CALVIN GIDDINGS

EXECUTIVE EDITOR

DEPARTMENT OF CHEMISTRY
UNIVERSITY OF UTAH
SALT LAKE CITY, UTAH

ELI GRUSHKA

GAS CHROMATOGRAPHY

DEPARTMENT OF CHEMISTRY
STATE UNIVERSITY OF NEW YORK AT BUFFALO
BUFFALO, NEW YORK

ROY A. KELLER

LIQUID CHROMATOGRAPHY

DEPARTMENT OF CHEMISTRY
STATE UNIVERSITY OF NEW YORK
COLLEGE AT FREDONIA
FREDONIA, NEW YORK

JACK CAZES

MACROMOLECULAR CHROMATOGRAPHY

WATERS ASSOCIATES, INC.
MILFORD, MASSACHUSETTS

CRC Press
Taylor & Francis Group
Boca Raton London New York

CRC Press is an imprint of the
Taylor & Francis Group, an **informa** business

First published 1975 by Marcel Dekker, Inc.

Published 2019 by CRC Press
Taylor & Francis Group
6000 Broken Sound Parkway NW, Suite 300
Boca Raton, FL 33487-2742

© 1975 by Taylor & Francis Group, LLC
CRC Press is an imprint of Taylor & Francis Group, an Informa business

First issued in paperback 2019

No claim to original U.S. Government works

ISBN-13: 978-0-367-45211-7 (pbk)
ISBN-13: 978-0-8247-6206-3 (hbk)

Visit the Taylor & Francis Web site at
http://www.taylorandfrancis.com

and the CRC Press Web site at
http://www.crcpress.com

LIBRARY OF CONGRESS CATALOG CARD NUMBER 65-27435

PREFACE

This volume of <u>Advances in Chromatography</u> marks a significant departure from earlier volumes in the series in editorial approach. Roy Keller, after shouldering for many years the major burden of general editorial work, has opted to restrict his efforts to a single field of chromatography: liquid chromatography. Accordingly, the editorial structure has been revised. We have created several "area editorships" to permit a greater editorial concentration in the major subfields of chromatography. We hope to generate a more meaningful coverage of each chromatographic area as the various area editors are able to focus their talents more specifically. The focus on liquid chromatography is provided by Roy Keller. Eli Grushka has taken responsibility for the broad and still active field of gas chromatography. Jack Cazes is focusing on large molecules; he is serving as area editor for macromolecular chromatography. Other area editors may be added in the future in order to keep in step with the ever-dynamic field of chromatography.

While emphases may shift, our goal here and in the future is the same goal we have pursued for eleven previous volumes of <u>Advances</u> beginning in 1965: to provide the community of chromatographers with stimulating, critical, readable, and relevant reviews of this broad and multifaceted subject. We welcome suggestions from readers to help us better reach these objectives.

J. Calvin Giddings
Executive Editor

CONTRIBUTORS TO VOLUME 12

PHYLLIS R. BROWN,* Section of Biochemical Pharmacology, Division of Biological and Medical Sciences, Brown University, Providence, Rhode Island

GARY J. FALLICK, Waters Associates, Inc., Milford, Massachusetts

ELI GRUSHKA, Department of Chemistry, State University of New York at Buffalo, Buffalo, New York

VIRGIL R. MAYNARD,† Department of Chemistry, State University of New York at Buffalo, Buffalo, New York

AKIRA NONAKA, Institute for Optical Research, Kyoiku University, Tokyo, Japan

LEON SEGAL, Southern Regional Research Laboratories, Agricultural Research Service, U.S. Department of Agriculture, New Orleans, Louisiana

JOSEPH SHERMA, Department of Chemistry, Lafayette College, Easton, Pennsylvania

DOUGLAS H. SMITH,‡ Hewlett-Packard Laboratories, Palo Alto, California

*Present address: Chemistry Department, University of Rhode Island, Kingston, Rhode Island

†Present address: Analytical Research and Services Laboratory, Central Research Laboratories, 3M Company, St. Paul, Minnesota

‡Present address: Hewlett-Packard, Avondale, Pennsylvania

CONTENTS

Chapter 1

THE USE OF HIGH-PRESSURE LIQUID CHROMATOGRAPHY
IN PHARMACOLOGY AND TOXICOLOGY   1

      Phyllis R. Brown

Chapter 2

CHROMATOGRAPHIC SEPARATION AND MOLECULAR-WEIGHT
DISTRIBUTIONS OF CELLULOSE AND ITS DERIVATIVES   31

Leon Segal

Chapter 3

PRACTICAL METHODS OF HIGH-SPEED
LIQUID CHROMATOGRAPHY   61

Gary J. Fallick

Chapter 4

MEASUREMENT OF DIFFUSION COEFFICIENTS BY GAS-
CHROMATOGRAPHY BROADENING TECHNIQUES: A REVIEW   99

Virgil R. Maynard and Eli Grushka

Chapter 5

GAS-CHROMATOGRAPHY ANALYSIS OF POLYCHLORINATED
BIPHENYLS AND OTHER NONPESTICIDE ORGANIC POLLUTANTS   141

Joseph Sherma

Chapter 6

HIGH-PERFORMANCE ELECTROMETER SYSTEMS FOR GAS
CHROMATOGRAPHY   177

Douglas H. Smith

Chapter 7

STEAM CARRIER GAS-SOLID CHROMATOGRAPHY   223

Akira Nonaka

CONTENTS OF OTHER VOLUMES

ADVANCES IN
CHROMATOGRAPHY
Volume 12

Chapter 1

THE USE OF HIGH-PRESSURE LIQUID
CHROMATOGRAPHY IN PHARMACOLOGY AND TOXICOLOGY

Phyllis R. Brown*

Section of Biochemical Pharmacology
Division of Biological and Medical Sciences
Brown University
Providence, Rhode Island

I. INTRODUCTION

Among the most difficult problems facing the researcher in chemistry and biology are the isolation, purification, and identification of compounds in a mixture. This is especially true in pharmacology and toxicology since

---

*Present address: Chemistry Department, University of Rhode Island, Kingston, Rhode Island

1

physiological fluids or cell extracts may contain hundreds of complex molecules with similar solubility properties. Some separation problems were solved by gas chromatography, others by paper and thin-layer chromatography; however, for many biologically active compounds these techniques were unsatisfactory. The development of high-pressure liquid chromatography (HPLC) has made possible the rapid analysis of nonvolatile, ionic, thermally labile compounds that were previously difficult or almost impossible to separate [1-3]. By the use of this method, molecular components in cells can be determined with high sensitivity, speed, accuracy, and resolution. The retention times are reproducible, and the instrument is versatile and simple to operate. Since the only major requirement is that the solutes be soluble in the mobile phase, a wide range of compounds can be analyzed rapidly and efficiently. This technique has been especially useful in the separation of drugs and their metabolites, and in the analysis of such normal constituents of cells as steroids and nucleotides in very small samples of cells.

Pharmacology is the science that deals with all aspects of drugs: their chemical composition, their biological action, and their therapeutic effects. It includes toxicology, the science that deals only with the poisonous effects of chemicals. A drug is defined here as any chemical compound that is used medically in the treatment, diagnosis, or prevention of disease or for the relief of pain. Drug investigations are complex since the nature, intensity, and duration of a drug's action are dependent on the combined effects of the absorption, distribution, biological transformation, and excretion of the drug.

The analysis of drugs and their metabolites in body fluids or tissues often presents technical difficulties. Usually only small samples are available. Speed and accuracy of analysis are demanded, and other substances may be present that interfere with the required analysis. Furthermore, many drugs are large complicated molecules that are changed on biotransformation into products that are more polar than the parent drug.

In pharmacology and toxicology three types of analyses are required:

1. The analysis of a drug or food itself to determine its composition and/or any decomposition products or impurities that are present

2. The determination of the concentration of a drug and/or its metabolites in physiological fluids or tissues

3. The analysis of cellular constituents to determine the effect of the drug on the naturally occurring cellular components

An example of the first category was the determination of the causative agent in an outbreak of cancer in turkey flocks in England. It was found that

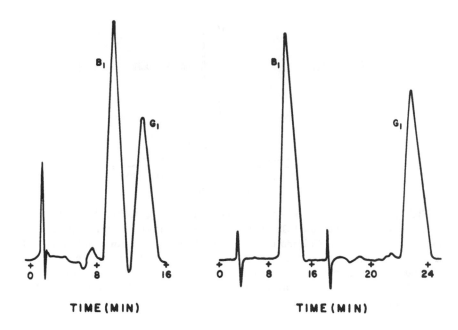

TIME (MIN)   TIME (MIN)

FIG. 1. Aflatoxins B_1 and G_1. Instrument: Perkin-Elmer liquid chromatograph. Column-packing material: Silosex adsorbent. Column dimensions: 3 mm x 50 cm. Solvent: 10:90 tetrahydrofuran-isopropyl ether. Flow rate: 0.82 ml/min; pressure: 250 psi. Detector: ultraviolet. Attenuation: 0.02 O.D. Courtesy of Ralph Conlon, Perkin-Elmer Corporation.

the carcinogen was the potent toxin aflatoxin, which is present in peanut meal contaminated with a fungus, Aspergillus flavus [4]. Although thin-layer chromatography was used to detect the deadly poison in this case, high-pressure liquid chromatography is now the method of choice because of the speed of analysis and ease in handling the samples. The separation of aflatoxins B_1 and G_1 is shown in Fig. 1.

An example of the second category is illustrated in Fig. 2. Anders and Latorre [5] showed that the glucuronide and sulfate metabolites of a drug in urine can be readily separated by high-pressure liquid chromatography.

The third category is illustrated in Fig. 3 by the work of Scholar, Brown, and Parks [6], who monitored with high-pressure liquid chromatography the synergistic effect of 6-methylmercaptopurine riboside on the adenine nucleotides of Sarcoma 180 cells.

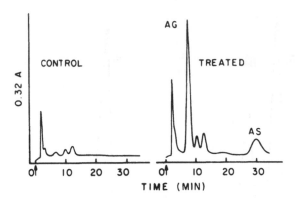

FIG. 2. Glucuronide and sulfate conjugates, metabolites in human urine. Instrument: Varian LCS 1000. Column-packing material: pellicular anion-exchange resin. Column dimensions: 250 mm x 1 mm. Solvent: 1.0 mM formic acid, pH 4.0; 1.0 mM formic acid, pH 4.0, containing 2.0 M KCl. Detector: ultraviolet. Temperature: 80°C. AG is glucuronide conjugate of acetanilide; AS is sulfate conjugate of acetanilide. Courtesy of M. W. Anders and J. Latorre.

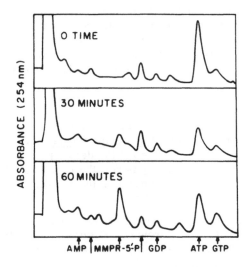

FIG. 3. Synergistic effect of 6-methylmercaptopurine riboside and 6-mercaptopurine on the adenine nucleotides of Sarcoma 180 cells. Instrument: Varian LCS 1000. Column-packing material: pellicular anion-exchange resin. Column dimensions: 3 m x 1 mm. Solvents: 0.015 M KH_2PO_4 and 0.25 M KH_2PO_4 in 2.2 M KCl. Flow rates: 12 and 6 ml/h. Temperature: 75°C. Detector: ultraviolet. Courtesy of E. M. Scholar et al.

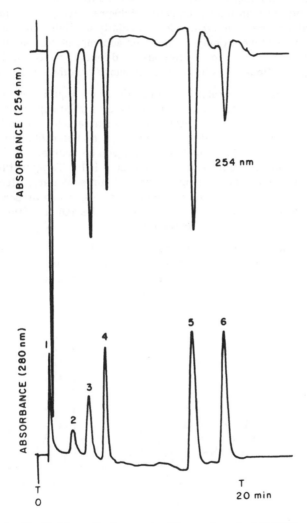

FIG. 4. Pesticides. Instrument: Chromatronix LC 3100. Column-
packing material: Vydac. Column dimensions: 2.1 x 250 mm. Solvents:
solvent A, isooctane; solvent B, chlorobutane. Flow rate: 2% B → 100% B
@ 5%/min, 5 ml/min. Detector: Model 230, ultraviolet. Sample size: 20 μl.
Peaks: 1, DDT; 2, Dyfonate; 3, methoxychlor; 4, parathion; 5, Imidan;
6, Sevin. Courtesy of Chromatronix, Inc.

The liquid-chromatography systems referred to throughout this chapter
are those that utilize high inlet pressure, high-pressure pumps, and sensitive low-dead-volume detectors. Commercially available instruments or
those built from components may be used for accomplishing the analyses
described. The chromatographic methods used include partition (liquid-
liquid), adsorption (liquid-solid), or ion-exchange chromatography.

The time required for analysis by high-pressure liquid chromatography
varies according to the separation; for example, a good separation of the
insecticides aldrin, lindane, and DDT (Fig. 4) can be achieved in less than
20 min. As shown in Figs. 5 and 6, respectively, a nucleotide profile of a
cell extract was achieved in about 1 h [7, 8], and an analysis of the carbohydrates in urine was accomplished in about 20 h [9]. Temperatures range
from ambient to 75°C, depending on the retention mechanism, eluents, and
type of compound to be separated. Solvents can be aqueous or nonaqueous,
depending on the type of column packing utilized, the nature of the separation,
and the kinds of compounds to be analyzed.

Good separations of various compounds depend on the temperature,
column dimensions, packing, concentration and pH of eluents, flow rates,
and type of elution. These conditions and the type of detector used will be
stated in the caption of a given chromatogram.

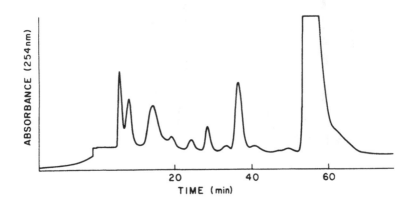

FIG. 5. Nucleotides in whole blood of normal adult (fasted). Instrument:
Varian LCS 1000. Column dimensions: 1 mm x 3 m. Column packing:
pellicular anion-exchange resin. Eluents: 0.015 M KH_2PO_4 and 0.25 M
KH_2PO_4 in 2.2 M KCl. Flow rates: 12 and 6 ml/h. Temperature: 75°C.
Detector: ultraviolet, 254 nm. Courtesy of A. Clifford and P. R. Brown,
unpublished data.

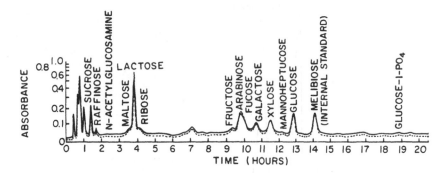

FIG. 6. Carbohydrates in the urine of a normal adult. Instrument: Carbohydrate Analyzer. Column dimensions: 150 × 0.22 cm i.d. Column packing: anion-exchange resin (Aminex A-27). Eluent: sodium tetraborate-boric acid buffer, pH 8.5; composition varies from 0.169 to 0.845 M. Temperature: 55°C. Detection: continuous colorimetric reaction of sulfuric acid and phenol with carbohydrates. Courtesy of C. D. Scott.

II. SAMPLE PREPARATION

Excellent progress has been made in the last 4 years in the application of high-pressure liquid chromatography in monitoring the disposition of drugs in man and animals. Good sample preparation is of the utmost importance to prevent sizable quantitative errors. Since errors can be cumulative in multistep preparation, it is important that the simplest and safest procedure be used. Examples of possible sources of errors are (1) sampling techniques; (2) sample preparation; (3) storage methods; (4) calculation errors; or (5) operator prejudice and/or carelessness.

The HPLC method provides a rapid, simple, sensitive technique for drug analyses that previously required long analysis time or multiple separation steps. For example, paper chromatography is the current U.S. Pharmacopeia procedure for the analysis of three sulfapyrimidine drugs that are commonly used in combination. The procedure requires 10 h to complete. With high-pressure liquid chromatography, Waters Associates, Inc., separated these compounds in 20 min (Fig. 7). Moreover, the concentration determinations are more accurate than those obtained with the paper-chromatography procedure.

For each class of drugs and their metabolites, specific extraction procedures must be determined. Procedures used in other chromatographic techniques, however, can be adapted to high-pressure liquid chromatography. For example, the combined use of Amberlite XAD-2 resin and thin-layer chromatography is currently being employed to screen such compounds as

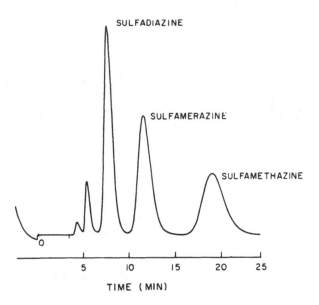

FIG. 7. Sulfa drugs. Instrument: Waters ALC 202. Column dimensions: 6 ft x 2.3 mm i.d. Column-packing material: WBAX-10. Solvent: water. Flow rate: 0.5 ml/min. Detector: ultraviolet. Courtesy of Waters Associates, Inc.

barbiturates, opium alkaloids, stimulants, and tranquilizers [10]. It is feasible to use the HPLC method in place of thin-layer chromatography with the XAD-2 resin to achieve these separations more rapidly and with better quantitation. It should be noted that speed and simplicity of preparation and stability of samples on storage are essential if the procedure is to be used clinically. In accordance with these criteria, Brown and Miech [11] worked out methods that can be used in the research and clinical laboratory for the rapid extraction of nucleotides prior to HPLC analysis.

III. STANDARDS

Chromatograms of standard compounds must be obtained for use as reference before applying HPLC techniques to cell extracts so that drug peaks may be readily identified. Examples of chromatograms of drugs are erythromycin, a broad-spectrum antibiotic (Fig. 8); vinblastine, a potent antineoplastic agent (Fig. 9); and the commonly used and abused drugs morphine and codeine (Fig. 10). With barbiturate poisoning on the increase, the rapid analysis of barbiturates is of vital importance. Standards of 12 barbiturates have been obtained using partition chromatography with chloroform as the eluent. Some of these standards are shown in Fig. 11. Often it is not the drug itself that causes toxicity but the impurities present. Therefore it is necessary to determine not only the kind and quality of barbiturate in a pill but also the impurities present. Analyzing for purity is also obviously important in industrial quality control.

There is great interest in the phenethylamines, which have profound effects on the central nervous system. A chromatogram of the separation of L-dopa, a drug used to treat Parkinson's disease, from its metabolite dopamine is shown in Fig. 12. The analysis of some other phenethylamines (methamphetamine, methoxyphenamine, and ephedrine) is shown in Fig. 13, and compounds such as ephedrine, theophylline, and phenobarbital, which are commonly present in asthma or hay-fever tablets, can readily be determined by HPLC (Fig. 14).

The separation of the analgetics aspirin and phenacetin from caffeine is demonstrated in Fig. 15. Benzoic acid was used as the internal standard.

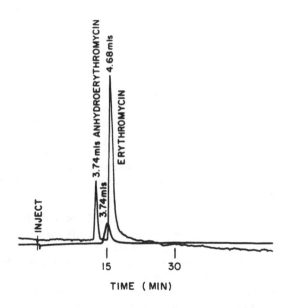

FIG. 8. Erythromycin. Instrument: Waters ALC-100. Column-packing material: Corasil II. Column dimensions: 3 ft × 2 mm i.d. Solvent: chloroform. Flow rate: 0.42 ml/min. Detectors: ultraviolet and refractive index. Courtesy of Waters Associates, Inc.

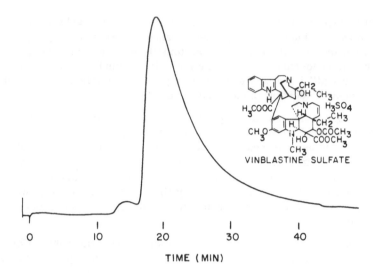

VINBLASTINE SULFATE

TIME (MIN)

FIG. 9. Vinblastine. Instrument: Waters ALC 202/401. Column-packing material: Corasil I coated with Duraphase-1-propylamine. Column dimensions: 2.3 mm × 4 ft. Solvents: 0.01 M sodium acetate and 0.05 M sodium sulfate. Flow rate: 0.37 ml/min. Detector: ultraviolet at 254 nm. Courtesy of Waters Associates, Inc.

Beyer [12] employed high-pressure liquid chromatography to provide a highly specific and practical method for the analysis of the sulfonylurea antidiabetic agents. Excellent HPLC separation of the benzodiazepines, another group of compounds of pharmacological interest, was obtained by Scott and Bonner [13]. The applications laboratories of the major liquid-chromatograph manufacturers now have standards available for many different types of drugs, such as diuretics, anticonvulsants, antibiotics, decongestants, and antineoplastic agents.

The ability to determine rapidly and accurately steroid levels in the physiological fluids of patients is becoming more important because of the use of steroids in the treatment of patients with such diseases as cancer, asthma, or arthritis. Although gas chromatography has been useful in the analysis of steroids, the HPLC method may prove to be more effective and efficient. In order to determine the feasibility of the analysis of steroids by this technique, Siggia and Dishman [14], Henry et al. [15], and the application laboratories of the various instrument companies determined the operating conditions necessary to separate many steroids. An illustration of the analysis of several steroids is shown in Fig. 16.

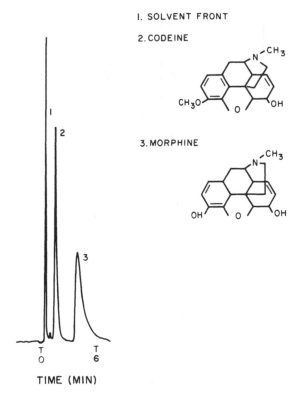

FIG. 10. Morphine and codeine. Instrument: Chromatronix. Column dimensions: 250 x 2.1 mm. Column packing: SS VYDAC. Solvents: 10% A, A = methylene chloride; 90% B, B =methylene chloride-methanol, 97:3. Flow rate: 1.6 ml/min; room temperature. Detector: Model 200 ultraviolet at 254 nm. Sample size: 20 μl. Sample concentration: 500 ppm. Peaks: 1, solvent front; 2, codeine; 3, morphine. Courtesy of Chromatronix, Inc.

The HPLC method is rapid and convenient for routinely analyzing nucleotides in physiological fluids and tissues [9]. It is especially useful in monitoring the metabolites of thiopurines, which show promise as anti-neoplastic and immunosuppressive agents. The nucleotides of thioanalogs of purines have longer retention times than their naturally occurring counter-parts and thus are cleanly separated. A chromatogram of some of the naturally occurring nucleotides is shown in Fig. 17, and one of the 5'-monophosphate of 6-methylmercaptopurine riboside as obtained by Brown [8] is shown in Fig. 18. The retention times of such metabolites as thiouric

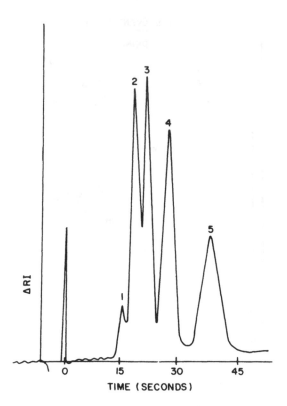

FIG. 11. Barbiturates. Instrument: Waters ALC-202. Column dimensions: 965 x 2 mm i.d. Column-packing material: 15% Carbowax 400 on Liqua-Chrom. Solvent: n-heptane-tetrahydrofuran, 90:10, v/v. Flow rate: 1.3 ml/min. Detector: ultraviolet detector at 254 and 280 nm. Peaks: 1, hexobarbital; 2, mephobarbital; 3, seco-barbital; 4, apobarbital; 5, barbital. Courtesy of Waters Associates, Inc.

acid and the 5′-mono-, 5′-di-, and 5′-triphosphate nucleotides of thio-guanosine in relationship to other nucleotides have been reported by Nelson and Parks [16]. The determination of these and other thiopurine metabolites under different elution conditions was described by Nelson et al. [17].

Wu and Siggia [18] also used high-pressure liquid chromatography in analyzing purine and strychnos alkaloids. They achieved fast separation of the pharmacologically important purines caffeine, theobromine, and theophylline (Fig. 19), and of the strychnos alkaloids brucine and strychnine by using liquid-liquid chromatography with a heptane-ethanol mobile phase.

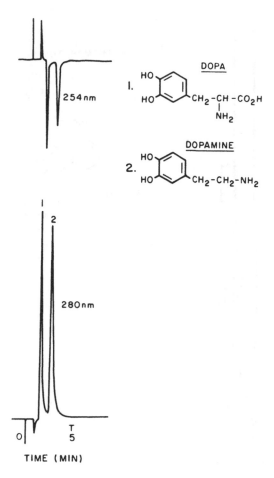

FIG. 12. L–Dopa and dopamine. Instrument: Chromatronix LC 3100. Column dimensions: 2.1 x 1000 mm. Column packing: VYDAC cation-exchange resin. Solvent: 0.1 M $NaNO_3$, pH 2.4. Flow rate: 1.2 ml/min at 1000 psi. Temperature: 60°C. Detector: Model 230 ultraviolet. Sample size: 20 μl. Sample concentration: 100-ppm solution. Peaks: 1, dopa; 2, dopamine. Courtesy of Chromatronix, Inc.

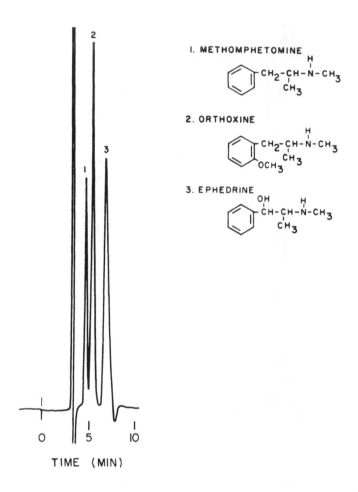

TIME (MIN)

FIG. 13. Phenethylamines. Instrument: Chromatronix LC 3100.
Column dimensions: 2 x 1500 mm. Column-packing material: Corasil
II. Solvent: 4:1 (chloraform-methanol) + 2 drops triethylamine and 1
drop 30% NaOH/l. Flow rate: 0.8 ml/min at 400 psi. Room temperature.
Detector: ultraviolet at 254 nm. Sample size: 20 μl. Sample concentration: about 50 to 100 ppm each. Peaks: 1, methamphetamine;
2, methoxyphenamine; 3, ephedrine. Courtesy of Chromatronix, Inc.

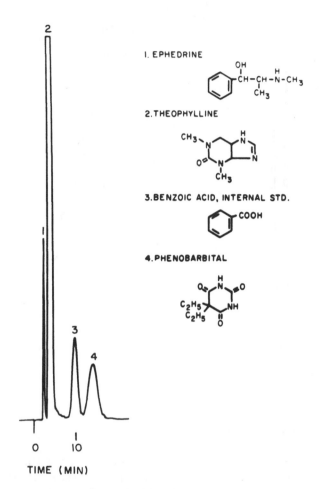

TIME (MIN)

FIG. 14. Asthma and hay-fever tablets. Instrument: Chromatronix LC 3100 UV. Column dimensions: 2 x 1500 mm. Column-packing material: SAX/ZIPAX. Solvent: 0.01 M NaNO₃, pH 5.7. Flow rate: 1 ml/min. Temperature: 37°C. Detector: ultraviolet at 254 nm. Sample size: 20 μl. Sample concentration: one tablet in 50 ml solvent. Peaks: 1, ephedrine; 2, theophylline; 3, benzoic acid (internal standard); 4, phenobarbital. Courtesy of Chromatronix, Inc.

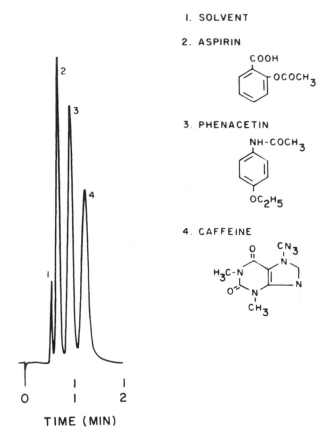

1. SOLVENT

2. ASPIRIN

3. PHENACETIN

4. CAFFEINE

FIG. 15. Analgesics. Instrument: Chromatronix LC 3100. Column dimensions: 2.4 x 1000 mm. Column-packing material: VYDAC cation-exchange resin. Solvent: 0.1 M NaCl, pH 2.5. Flow rate: 3.5 ml/min at 1800 psi. Temperature: 60°C. Detector: Model 230 at 254 nm. Attenuation: 16x. Chart speed: 1.0 in./min. Sample size: 2.0 μl. Peaks: 1, solvent; 2, aspirin; 3, phenacetin; 4, caffeine. Courtesy of Chromatronix, Inc.

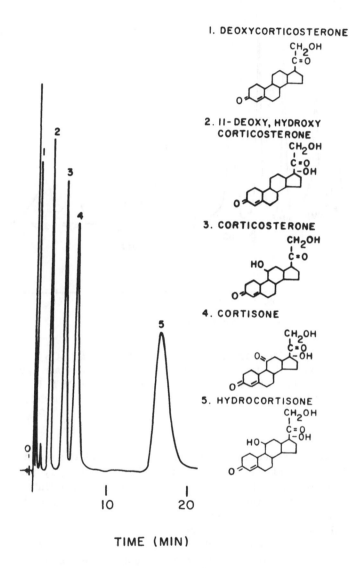

TIME (MIN)

FIG. 16. Steroids. Instrument: Chromatronix LC 566. Column dimensions: 2 × 500 mm. Column-packing material: VYDAC cation-exchange resin. Solvent: 7:3 isooctane–chloroform, with 3% methanol. Flow rate: 1.2 ml/min at 300 psi. Room temperature. Detector: ultraviolet Model 200 at 259 nm. Sample size: 30 μl. Sample concentration: 50 to 80 μg/ml. Peaks: 1, deoxycorticosterone; 2, 11-deoxyhydroxycorticosterone; 3, corticosterone; 4, cortisone; 5, hydroxycortisone. Courtesy of Chromatronix, Inc.

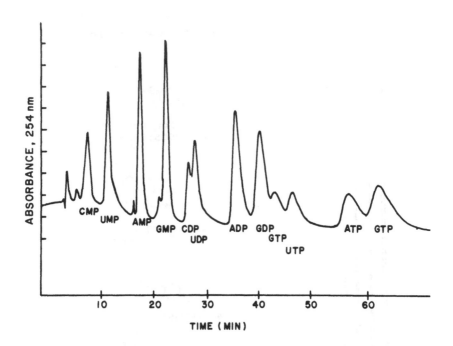

FIG. 17. Nucleotide standards. Instrument: Varian LCS 1000.
Column dimensions: 1 mm × 3 m. Column-packing material: pellicular
anion-exchange resin. Eluents: 0.015 M KH_2PO_4 and 0.25 M KH_2PO_4
in 2.2 M KCl. Flow rates: 12 and 6 ml/h. Detector: ultraviolet at
254 nm.

In order to ensure accuracy of results, factors affecting quantitation
were examined. For all compounds used regularly, Beer's law plots
should be determined. A modified Beer's law plot of the concentration of
GTP versus height times width at half-height of the GTP peak is shown in
Fig. 20. If, as is the case with many drugs, an ultraviolet monitor is
used, it should be noted that extinction coefficients vary from compound
to compound at the given wavelength. Therefore, for accurate concentrations, calibration factors for each compound are necessary. In the
analysis of nucleotides it was also found that flow rate affected concentration determinations but that the slope of the linear concentration
gradient was not involved in the calculation of concentrations [19].

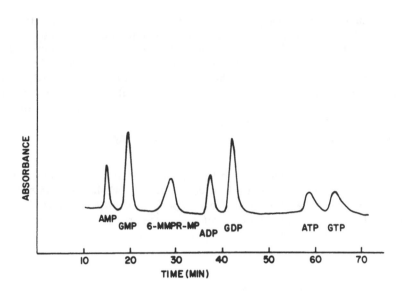

FIG. 18. 6-Methylmercaptopurine riboside 5'-monophosphate and adenine and guanine ribonucleotides. Instrument: Varian LCS 1000. Column dimensions: 3 m x 1 mm i.d. Column-packing material: Varian LFS pellicular anion-exchange resin. Solvents: Low-concentration buffer, 0.015 M KH_2PO_4, pH 4.5; high-concentration buffer, 0.25 M KH_2PO_4 in 2.2 M KCl, pH 4.5. Flow rate: 12 ml/h. Temperature: 70°C. Detector: ultraviolet at 254 nm.

IV. BASE-LINE STUDIES

Several groups of investigators have published HPLC base-line studies of the nucleotide patterns of various normal cells, such as the formed elements of human blood [20], rat brain [21], rat liver [22], and human urine [9, 23, 24]. In determining the effect of a drug on cell growth it is advisable to monitor the effect of the drug on the free nucleotide pools on the cells as well as the concentrations of the drug and its metabolites. It has been found by Horvath et al. [7] and Brown [8] that the nucleotide profiles of cell extracts can be obtained rapidly and easily and that the results were quantitatively reproducible. An example of the nucleotide profile of rat liver is shown in Fig. 21. Moreover, the ribosides and bases can be readily analyzed by this technique [25].

Since high-pressure liquid chromatography has unique potential for clinical use in monitoring changes in nucleotide pools caused by disease states or during drug therapy, nucleotide patterns of the acid-soluble extracts of the formed elements in normal human peripheral blood were determined [20]. The erythrocytes, leukocytes, and platelets had characteristic and reproducible nucleotide patterns. There was little variation in the concentrations of the major nucleotides (ATP, ADP, GDP, UTP, etc.) in each fraction of the blood from normal adults. Because of the high resolution and sensitivity afforded, this technique also has the advantage of being able to detect small changes in nucleotide patterns caused by disease or drug therapy — changes that may not be detected by other methods.

To establish which species are the best models in pharmacology studies, the species differences in the nucleotide patterns of whole blood were determined [26]. It was found that each species exhibited a characteristic variation in the nucleotide profile, but within the species the chromatograms were very reproducible.

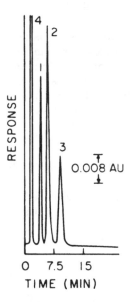

FIG. 19. Purine alkaloids. Column dimensions: 1 m x 1 mm. Column-packing material: Corasil II coated with 1.1% Poly G-300. Eluent: n-heptane-ethanol (100:10, v/v). Flow rate: 0.27 ml/min. Detector: ultraviolet at 270 nm. Peaks: 1, caffeine; 2, theophylline; 3, theobromine; 4, solvent front. Courtesy of C. -Y. Wu and S. Siggia.

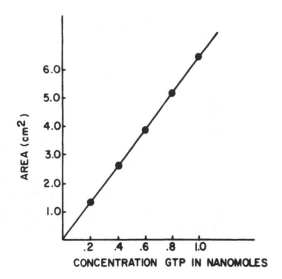

FIG. 20. Modified Beer's law plot of concentration of GTP versus height x width at one-half the height of peak areas.

Comprehensive base-line studies have been carried out on urine at the Oak Ridge National Laboratory [9, 23, 24]. There are more than 300 molecular constituents in normal human urine, and at least 140 chromatographic peaks have been resolved. This information is being used to evaluate body function and to detect pathological conditions. Since excretion is an important function in the study of drug metabolism, this basic work is vital to pharmacology studies so that drugs and drug metabolites can be determined in relation to the naturally occurring constituents.

V. PEAK IDENTIFICATION

In analyzing cell extracts by the HPLC technique, the peaks in the chromatograms of these extracts must be identified. Peaks can be identified by retention times, classical chemical and spectrophotometric methods, derivatization, and the enzymatic-peak-shift method [8]. The enzymatic-peak-shift technique, which utilizes the specificity of enzyme reactions with a functional group, a compound, or a class of compounds, has proved to be of great value. It was especially useful in the characterization of nucleotides of cell extracts; not only is the identity of the reactant verified but also that of the product formed. An example is the reaction of ADP + GDP with phosphoenolpyruvate (PEP) and pyruvate kinase (PK),

$$\text{GDP} + \text{ADP} + 2\text{PEP} \xrightarrow{\text{PK}} \text{GTP} + \text{ATP} + 2 \text{ pyruvate}$$

in which the ADP + GDP peaks disappear and the ATP and GTP peaks increase proportionally (Fig. 22). Moreover, these reactions are helpful in clarifying a chromatogram. If one nucleotide is present in large quantity, it may hide the presence of a small quantity of another nucleotide. Therefore the enzymatic-peak-shift method can be used in the quantitation of a hidden peak or the determination of the shape of a peak that otherwise might be seen only as a shoulder. Besides specificity of the reaction, a major requirement for this technique is that the reagents be available, inexpensive, and have retention times that differ from both product and reactant if they adsorb in the ultraviolet at interfering wavelengths.

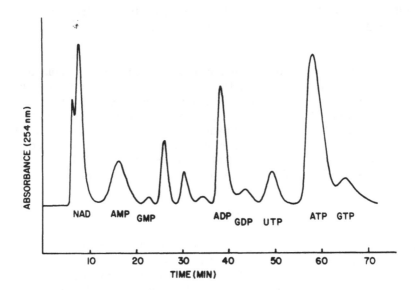

FIG. 21. Nucleotide profile of rat liver. Instrument: Varian LCS 1000. Column dimensions: 3 m x 1 mm i. d. Column-packing material: Varian LFS pellicular anion-exchange resin. Solvents: low-concentration buffer, 0.015 M KH_2PO_4, pH 4.5; high-concentration buffer, 0.25 M KH_2PO_4 in 2.2 M KCl, pH 4.5. Flow rates: column flow rate, 12 ml/h; gradient flow rate, 6 ml/h. Temperature: 70°C. Detector: ultraviolet at 254 nm. From unpublished work by A. Clifford and P. R. Brown.

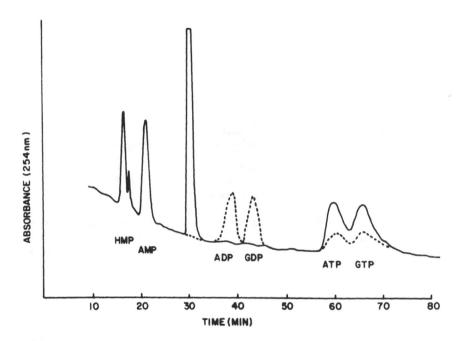

FIG. 22. Enzymatic peak shift. Instrument: Varian LCS 1000. Column dimensions: 3 m x 1 mm i.d. Column packing material: Varian LFS pellicular anion-exchange resin. Solvent: low-concentration eluent, 0.015 M KH_2PO_4, pH 4.5; high-concentration eluent, 0.25 M KH_2PO_4 in 2.2 M KH_2PO_4, pH 4.5. Flow rates: column, 12 ml/h; gradient, 6 ml/h. Temperature: 70°C. Detector: ultraviolet at 254 nm. The reaction is that of ADP and GTP with phosphoenolpyruvate in the presence of pyruvate kinase to give ATP, GTP, and pyruvate. The broken line is the chromatogram before the reaction and the solid line is after the reaction.

VI. PHARMACOLOGICAL STUDIES

The HPLC method has proved valuable in separating drugs from their metabolites in physiological fluids and cell extracts. For example, in the treatment of epilepsy, phenobarbital and diphenylhydantoin are often prescribed. Anders and Latorre [27] showed the ease of analyzing by high-pressure liquid chromatography these compounds and their metabolites, the hydroxylated compounds. The technique can be used in determining the chemistry of metabolites, the concentrations of metabolites formed, and the distribution of the drugs and their metabolites in physiological fluids. It can also be applied to the studies of synergistic effects of drugs, the species differences in the metabolism of drugs, and the effect of various factors (e.g., diet, age, sex, and health) on drug metabolism.

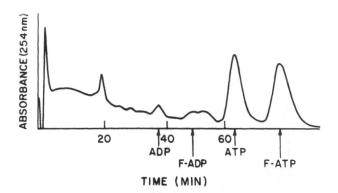

FIG. 23. Nucleotides in human erythrocytes incubated with 2-fluoro-
adenosine. Instrument: Varian LCS 1000. Column dimensions: 1 mm x
3 m. Column-packing material: pellicular anion-exchange resin. Eluents:
0.015 M KH_2PO_4 and 0.25 M KH_2PO_4 in 2.2 M KCl. Flow rates: 12 and
6 ml/h. Temperature: 75°C. Detector: ultraviolet at 254 nm.

The metabolic pathways of purine and purine nucleoside analogs have
been investigated in two model systems: a normal system (human erythrocytes) [28] and a tumor line (Sarcoma 180 mouse ascites cells) [6]. In the
study of normal cells, human erythrocytes are well suited for the investigation of certain facets of purine nucleotide metabolism because the cells
have no de novo pathway and do not synthesize DNA or RNA. Therefore the
metabolic pathways of purine analogs could be monitored without interference
from competing reactions. The incorporation of purine ribonucleoside
analogs was examined by high-pressure liquid chromatography along with
the effect of these drugs on the naturally occurring nucleotides. It was found
that 2-fluoroadenosine, tubercidin, and toyocomycin are readily phosphor-
ylated to the triphosphate level. The 6-sulfur analogs, however, form only
the monophosphate nucleotides. In the erythrocytes the drugs under
investigation had no effect on the naturally occurring nucleotide pools. An
example of a chromatogram of the nucleotides in human erythrocytes
incubated with 2-fluoroadenosine is shown in Fig. 23.

The applications of high-pressure liquid chromatography to pharmacological studies were also demonstrated in the investigations of the effect
of thiopurines on Sarcoma 180 mouse ascites cells. Although the tumor
cells have more complicated metabolic pathways than do human erythrocytes,
in metabolic investigations this cell line is a good model system for neo-
plastic disorders. Both in vitro and in vivo studies of the synergistic effect
of mercaptopurine analogs were carried out with Sarcoma 180 cells. When
both 6-methylmercaptopurine riboside and 6-mercaptopurine were incubated
with these tumor cells, there was a marked synergistic effect, which caused

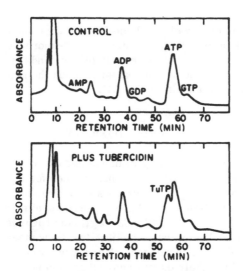

FIG. 24. Nucleotides in <u>Schistosoma</u> <u>mansoni</u> incubated with tubercidin. Instrument: Varian LCS 1000. Column dimensions: 1 mm x 3 m. Column-packing material: pellicular anion-exchange resin. Eluents: 0.015 M KH_2PO_4 and 0.25 M KH_2PO_4 in 2.2 M KCl. Flow rates: 12 and 6 ml/h. Temperature: $75^\circ C$. Detector: ultraviolet at 254 nm. Courtesy of R. J. Stegman et al.

a decrease in the cell's adenine nucleotide concentration (see Fig. 3) [6]. The biochemical mechanism was also investigated for the synergism of 6-thioguanine and 6-methylmercaptopurine riboside using the same cell line [16]. For these studies the HPLC method was valuable because it was possible to monitor the analog-nucleotide formation simultaneously with the effect of drugs on the naturally occurring nucleotide pools.

In conjunction with other techniques, high-pressure liquid chromatography has been used effectively in the search for a cure or control of the parasitic disease schistosomiasis. The purine metabolism of the causative agent, the blood fluke <u>Schistosoma</u> <u>mansoni</u>, was investigated, and it was found that schistosomes, like human erythrocytes, do not have a purine "de novo" synthesis pathway and therefore depend on preformed purines for their RNA and DNA [29]. In this investigation a variety of techniques were used: enzymatic assays, HPLC analysis of nucleotide pools, radioisotope studies, and paper chromatography. It has been found that tubercidin, toyocomycin, and 2-fluoroadenosine have possible chemotherapeutic value [30, 31], and promising chemotherapeutic results were obtained when tubercidin-loaded erythrocytes were infused into mice bearing <u>Sch</u>. <u>mansoni</u> [32]. Therefore the incorporation of these adenosine

analogs into the nucleotide pools of these blood flukes was examined [33].
Current HPLC work showed that tubercidin and 2-fluoroadenosine were
converted into the analog triphosphate nucleotides, and the addition of
tubercidin to intact schistosomes appeared to block the synthesis of
adenine nucleotides from adenine (Fig. 24).

VII. DRUG TOXICITY

Today, more than ever, the toxicity of drugs and other chemicals is of
real concern. Better technology is urgently needed for drug analyses in
pharmacology and toxicology laboratories so that drug levels in man can be
monitored quickly and efficiently. Since all drugs are toxic in overdose,
the increased availability of drugs leads to greater hazards of toxicity.
With the easier availability of drugs medically, the number of accidental
poisonings and suicides has increased. Moreover, people are exposed to
many toxic chemicals in their daily lives. It is also becoming increasingly
evident that various pesticides are toxic not only to insects but also to
humans, birds, and animals. Therefore a rapid, sensitive, and accurate
technique for analyzing pesticide composition and residues is needed. It
has been found that excellent separations of many insecticides, herbicides,
and larvacides (and their metabolites) can be achieved by high-pressure
liquid chromatography. An example of a good separation is that of
aldrin from DDT and p,p'-DDT (Fig. 25). The technique has advantages
over other methods previously used in that it is rapid and the nonpesticide
peaks do not usually interfere with the peak quantitation.

VIII. DRUG RESISTANCE

Although the HPLC technique has not yet been used in studies of drug
resistance, it has great potential for application to this problem. Drug
resistance refers to the state of insensitivity or decreased sensitivity to a
particular drug. In the clinical use of drugs to treat disease, organisms or
cells may become resistant to a drug that ordinarily causes growth
inhibition or cell death. There are a number of possible mechanisms of
drug resistance. Drug resistance can result from decreased intracellular
drug levels, decreased conversion of a drug to a more active metabolite,
or increased destruction of a drug. Changes in drug or drug metabolite
levels, which may indicate the mechanism of drug resistance, can be
readily monitored by high-pressure liquid chromatography. For example,
it is known that resistance develops to chloroquine, a drug used in the
treatment of malaria. In experiments in mice using Plasmodium berghei,
it was found that the development of resistance was related to the decreased
concentration of chloroquine in the plasmodium. This indicates that the
mechanism of resistance is the impeded uptake of the drug by the organism
[34, 35]. Although the HPLC method was not used for these particular
experiments, it can readily be adapted for use in such studies. An example

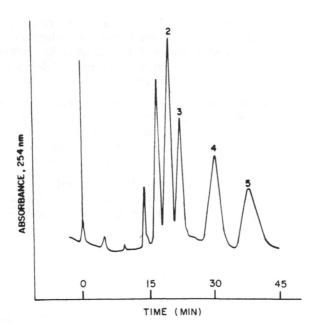

FIG. 25. Insecticides. Instrument: Waters ALC-201. Column
dimensions: 50 cm × 2.3 mm i.d. Column-packing material: Corasil I.
Solvent: n-hexane. Flow rate: 3.0 ml/min. Peaks: 1, aldrin impurity;
2, aldrin; 3, p,p'-DDT; 4, DDT; 5, lindane. Courtesy of Waters
Associates, Inc.

of another mechanism of resistance, the decreased conversion of the drug to
a more active metabolite, is the resistance developed by a <u>Salmonella</u> species
to 2,6-diaminopurine [36]. Since the drug forms a nucleotide on metabolism, the development of this resistance could easily be monitored by
high-pressure liquid chromatography by following not only the drug concentrations but also the formation of the metabolite, the nucleotide of this
purine.

IX. LIMITATIONS

The HPLC method has limitations that prevent its universal use in research
and clinical laboratories. The most obvious limitation is the newness of
the instrumentation and the lag experienced in adapting this technique to
biomedical research. Literature has not yet accumulated. Therefore the
best experimental conditions for the analysis of many compounds have not
been determined, and for each new group of compounds experimental
conditions must be optimized.

The most sensitive detector available today is the micro ultraviolet detector. It limits the use of high-pressure liquid chromatography because some drugs do not contain a chromophore that absorbs in the ultraviolet. Therefore a universal detector with the sensitivity and reliability of the ultraviolet detector is the ideal model sought.

Since the column is the heart of a chromatograph, the production of long-lasting packings for all types of separation is a necessity if the high-pressure liquid chromatograph is to gain general acceptance. The major difficulty has been in producing, from batch to batch, stable packings that will give reproducible retention times and good resolution of peaks.

For the HPLC technique to be practical in the clinical laboratory, the analyses must be rapid and the procedures automated. The instrument must be rugged and simple to operate. Furthermore, it must be easily serviced so that it can run continuously.

X. CONCLUSIONS

At present the high-pressure liquid chromatograph is used mainly in pharmacology research laboratories. It is predicted, however, that this instrument will eventually be standard equipment in the clinical and toxicology laboratory because of its sensitivity, versatility, and potential for accurately analyzing compounds of clinical and toxicological significance that are difficult to analyze by other methods. Furthermore, this instrument, if properly developed, will be an important clinical tool for monitoring drug therapy and for detecting drug toxicity and resistance.

ACKNOWLEDGMENT

This work has been supported by a grant (No. 16538) from the U.S. Public Health Service.

REFERENCES

1. N. Hadden, F. Baumann, F. MacDonald, M. Munk, R. Stevenson, D. Gere, and F. Zamaroni, Basic Liquid Chromatography, Varian Aerograph, Walnut Creek, Calif., 1971.

2. J. J. Kirkland, ed., Modern Practice of Liquid Chromatography, Wiley-Interscience, New York, 1971.

3. P. R. Brown, High Pressure Liquid Chromatography, Biochemical and Biomedical Applications, Academic Press, New York, 1973.

4. H. F. Kraybill and M. B. Shimkin, Adv. Cancer Res., 8, 191 (1964).

5. M. W. Anders and J. Latorre, J. Chromatogr., 55, 400 (1971).

6. E. M. Scholar, P. R. Brown, and R. E. Parks, Jr., Cancer Res., 32, 259 (1972).

7. C. G. Horvath, B. A. Preiss, and S. R. Lipsky, Anal. Chem., 41, 1227 (1969).

8. P. R. Brown, J. Chromatogr., 52, 272 (1970).

9. C. D. Scott, R. L. Jolley, W. W. Pitt, and W. F. Johnson, Am. J. Clin. Pathol., 53, 701 (1970).

10. J. M. Fujimotoa and S. H. Wang, Toxicol. Appl. Pharmacol., 6, 86 (1970).

11. P. R. Brown and R. P. Miech, Anal. Chem., 44, 1072 (1972).

12. W. F. Beyer, Anal. Chem., 44, 1313 (1972).

13. C. G. Scott and P. J. Bonner, J. Chromatogr. Sci., 8. 446 (1970).

14. S. Siggia and R. A. Dishman, Anal. Chem., 42, 1223 (1970).

15. R. A. Henry, J. A. Schmitt, and J. F. Diechman, J. Chromatogr. Sci., 9, 513 (1971).

16. J. A. Nelson and R. E. Parks, Jr., Cancer Res., 32, 2034 (1972).

17. D. J. Nelson, C. J. L. Bugge, H. C. Krasny, and T. P. Zimmerman, J. Chromatogr., 77, 181 (1973).

18. C. -Y. Wu and S. Siggia, Anal. Chem., 44, 1499 (1972).

19. P. R. Brown, J. Chromatogr., 57, 383 (1971).

20. E. M. Scholar, P. R. Brown, R. E. Parks, Jr., and P. Calabresi, Blood, 41, 927 (1973).

21. H. W. Schmukler, J. Chromatogr. Sci., 10, 38 (1972).

22. A. J. Clifford, J. A. Riumallo, B. S. Baliga, H. N. Munro, and P. R. Brown, Biochim. Biophys. Acta, 277, 443 (1972).

23. C. D. Scott, J. E. Attril, and N. G. Anderson, Proc. Soc. Exp. Biol. Med., 125, 181 (1967).

24. C. D. Scott, Clin. Chem., 14, 521 (1968).

25. C. G. Horvath and S. R. Lipsky, Anal. Chem., 41, 1227 (1969).

26. P. R. Brown, R. P. Agarwal, J. Gell, and R. E. Parks, Jr., Comp. Biochem. Physiol., 43B, 891 (1972).

27. M. W. Anders and J. Latorre, Anal. Chem., 42, 1430 (1970).

28. R. E. Parks, Jr., and P. R. Brown, Biochemistry, 12, 3294 (1973).

29. A. W. Senft, R. P. Miech, P. R. Brown, and D. Senft, Int. J. Parasitol., 2, 249 (1972).

30. C. G. Smith, L. M. Reineki, M. R. Burch, A. M. Shefner, and E. M. Murhead, Cancer Res., 30, 69 (1970).

31. J. A. Montgomery and K. Hewson, J. Am. Chem. Soc., 82, 463 (1960).

32. J. J. Jaffee, E. Meymarian, and H. M. Doremus, Nature, 230, 308 (1971).

33. R. J. Stegman, A. W. Senft, P. R. Brown, and R. E. Parks, Jr., Biochem. Pharmacol., 22, 459 (1973).

34. P. B. Macomber, R. L. O'Brien, and E. H. Hahn, Science, 152, 1374 (1966).

35. A. V. S. DeReuck and M. P. Cameron, eds., Ciba Foundation Study Group No. 13, Little, Brown and Co., Boston, 1962.

36. G. P. Kalle and J. S. Gots, Science, 142, 680 (1963).

Chapter 2

CHROMATOGRAPHIC SEPARATION AND MOLECULAR-WEIGHT
DISTRIBUTIONS OF CELLULOSE AND ITS DERIVATIVES

Leon Segal

Southern Regional Research Laboratories
Agricultural Research Service
U.S. Department of Agriculture
New Orleans, Louisiana

I. INTRODUCTION

Cellulose, the naturally occurring polymer of D-glucose, is of great technical and economic importance. Found in cotton, flax, ramie, hemp, and wood pulp — the celluloses of commerce — the polymer is familiarly handled in the form of fabrics, paper, rope, and the like. For many uses (e. g., fabric, film, gunpowder, and fabricated objects) cellulose is converted into chemical derivatives, such as cellulose acetate, cellulose xanthate, cellulose nitrate, methyl and ethyl celluloses, as well as other esters and ethers. The chemical and physical properties of cellulose and its derivatives, as well as chemical treatments applied to the polymer, are thoroughly covered in the vast literature on cellulose.

Like any other polymer, cellulose is polymolecular: it is composed of molecular chains of various lengths in a continuous distribution rather than molecules of single size or weight. The chain length, or degree of polymerization (DP), of a cellulose molecule, which is fixed by the number of anhydroglucose units connected together to form the chain molecule, is sensitive to many factors and varies from one cellulose to another. Because of the continuous distribution of chain lengths, the measured DP of a sample is an average value that varies according to the method of measurement. Thus there is the number-average DP ($\overline{DP}_n$) from osmometry, the weight-average DP ($\overline{DP}_w$) from ultracentrifugation, the viscosity-average DP ($\overline{DP}_v$) from viscometry, as well as the z average, $\overline{DP}_z$, and the z + 1 average, $\overline{DP}_{z+1}$. The $\overline{DP}_z$ and $\overline{DP}_{z+1}$ averages are calculated from $\overline{DP}_w$.

The ratio $\overline{DP}_w/\overline{DP}_n$ is called the polymolecularity ratio and theoretically is equal to unity when all of the molecular chains in a sample are of the same length. This ratio increases as the range of chain lengths increases, that is, as the molecular-weight distribution (MWD) or DP distribution broadens. However, use of this ratio as an index of DP distribution is misleading because of the variability hidden in the average values from which it is calculated. A more nearly true distribution is obtained by polymer fractionation — separation of the bulk polymer into fractions with rather narrow ranges of chain length.

Chromatography has been firmly established as an effective means of separating the components of a mixture. Numerous variations in methodology have made application of chromatographic separation extremely versatile and suitable for use in some in rather specific problem areas. An excellent, brief, and easily readable survey of the principal methods of chromatography has been written by Cates and Guion [1] with emphasis on textile applications. Cates and Guion, however, make only one reference to the application of a chromatographic procedure to a cellulosic material — pyrolysis-gas chromatography. In this case chromatography was not used for fractionation but only as a means of distinguishing between cellulose triacetate and a cellulose acetate of a lower degree of substitution. This paucity of references seems surprising because cellulose acetate was used by Mark and Saito [2] in 1936 in the first published report on the fractionation of high polymers by chromatographic methods. They demonstrated that cellulose acetate was selectively adsorbed on charcoal on a molecular-weight basis. The adsorption of cellulose acetates on charcoal was also studied by Levi and Giera [3]. However, it appeared from these studies that charcoal was not likely to be very useful in elution chromatography.

Years later, in 1949, an attempt was made to quantitatively fractionate cellulose nitrate by elution chromatography using a column of activated carbon and Supercel [4], but the results indicated that many problems had to be resolved before this technique could be successful. Partition chromatography, using swollen cellulose triacetate as the adsorbent and methyl acetate-water as cosolvent, was applied to cellulose nitrate in 1950 [5], but this too was fraught with problems. Factors causing trouble included a low rate of adsorption when the molecular weight was high, decreasing adsorption with increasing molecular weight, increasing adsorbability with increasing nitrogen content, and a shortage of suitable solvent systems. A demonstration that elution chromatography could possibly be used to fractionate cellulose nitrate developed from the experiments with partition chromatography [6]. Suitably prepared starch was the adsorbent, and the developer was a series of cosolvents of changing composition. Limitations of this procedure were (a) that the degree of nitration had to be held very constant, (b) that separation had to be carried out in a low concentration range, (c) that the developer or solvent system was not a simple one, and (d) that the cellulose nitrate was of quite low DP.

Cellulose itself is a very poor subject for fractionation by chromatographic methods. This is reflected, perhaps, in the absolute lack of published material on the subject prior to 1967. The problem is that cellulose is insoluble in solvents suitable for the usual chromatographic procedures. It is not soluble in water or in organic solvents. The cellulose molecule is brought into solution by metal chelate complexes dissolved in aqueous alkaline solutions. The principal, important solvents for cellulose are cupric ammonium hydroxide or cuprammonium (cuam), cupric ethyl-enediamine hydroxide or cupriethylenediamine (cuen), cadmium ethylene-diamine hydroxide (cadoxen), and iron-sodium tartrate (FeTNa). Under special conditions certain mineral acids at very specific concentrations and temperatures and certain salt solutions will dissolve some celluloses, but these solvents are special cases and unimportant for chromatographic considerations.

Not until the early 1960s were chromatographic procedures successfully used to fractionate cellulose and its derivatives. The two newer and less known procedures used by polymer chemists were "precipitation chromatography" and gel-permeation chromatography. Of these, the latter

technique rapidly eclipsed the former, for reasons that will become
obvious.

II. PRECIPITATION CHROMATOGRAPHY

A. The Method

Precipitation chromatography, a chromatographic procedure for the
fractionation of high polymers according to molecular weight, has been
reviewed very ably by Porter and Johnson [7]. The purist, however, may
protest that this procedure is not truly chromatography because the
mechanism of the separation is not that on which chromatography is
customarily based. Nonetheless, if one agrees with the definition that
"chromatography is a physical method of separation in which the components
to be separated are distributed between two phases, one of the phases
constituting a stationary bed of large surface area, the other being a fluid
that percolates through the stationary bed" [8], then precipitation chromatography is, indeed, chromatography.

The operations required to achieve separation are very similar to those
of elution or column chromatography. The elements of the apparatus used
for this fractionation are (a) a mixing device to provide a solvent-nonsolvent
mixture with a smoothly and continuously varying composition (a solvent
gradient), (b) a column packed with an inert material in a form having a
large surface area (usually 40- to 70-μ glass beads), (c) heaters around
the column to provide a temperature gradient along the column (warmer
at the top, cooler at the bottom), and (d) a sample collector. The polymer
sample is coated onto a small quantity of the glass beads by evaporation of
a solution prepared from the good solvent of the solvent pair; the coated
beads are then placed as a slurry on top of the packed column, using the
poor solvent to make up the slurry. The solvent-nonsolvent mixture is
introduced onto the coated beads, dissolving some of the polymer. The
solution travels downward, where the cooler portions of the column
precipitate the dissolved polymer onto the bead packing. There is a
progression of steps involving redissolving and reprecipitation as solvent
gradient and column temperature interact to bring about molecular separation. Fresh polymer dissolves at the top of the column as the solvent-
nonsolvent ratio changes. The above are the essentials of the Baker-
Williams [9] procedure. The shorter chain lengths appear first in the
eluate, with the longer lengths appearing progressively as the fractionation
proceeds. The molecular-weight distribution of the sample is obtained by
determining the amount of polymer in each fraction and its molecular
weight, then constructing integral and differential distribution curves from
these data. The greater the number of fractions collected and the narrower
the cut, the better the definition of the distribution curve.

B. Extent of Application

In spite of the general extensive use of precipitation chromatography by polymer chemists, its successful application has not been matched in the field of cellulose chemistry. There is very little to be found on this subject in the literature on cellulose derivatives. The chromatographic separation of cellulose itself, of course, cannot be achieved in this manner because of the limitations placed by the solvent requirements.

Heyemann [10] developed a column that operated satisfactorily with cellulose tricarbanilate. The usual combination of solvent gradient and temperature gradient was used for the column, but successful separation was dependent on the column packing. The usual glass beads of about 100 μ caused degradation of the cellulose chain after the polymer had resided a short time in the column; this was most noticeable with cellulose acetate and cellulose nitrate. The adverse effect was attributed to the high alkalinity of the glass. Fine sand was a poor packing material because of channeling; a Celite column packed too densely because the particle size of Celite is so small. Packing problems were surmounted by use of prepolymer polyester beads that had been extracted with solvent until the eluate was free of polymer (detected by examination of the eluate under ultraviolet light). Use of the carbanilate derivative instead of the nitrate for column fractionation was preferred for the following reasons: (a) it is more stable in the column, (b) it is soluble in a wider variety of suitable solvents, (c) it can be regenerated to cellulose with a minimum of degradation, and (d) it presents fewer problems in the preparation of the trisubstituted derivative.

III. GEL-PERMEATION CHROMATOGRAPHY

A. The Method

Gel-permeation chromatography (GPC) is widely used now for fractionating and obtaining the molecular-weight distributions of homopolymers. Although it was not considered a chromatographic procedure initially, the technique clearly falls within the definition of chromatography given by Chatterjee and Schwenker [8]. A fixed, small volume of a polymer solution is passed over porous media — either a swollen gel of a high polymer, porous glass, or silica beads — packed into columns. The solution is eluted by means of pure solvent, whereby separation of the molecular mixture takes place as a function of the pore size of the column-packing material. In this column fractionation technique the larger molecules elute first from the column, followed by the smaller ones — a direct reversal of that which occurs in precipitation chromatography. This sequence is explained by considering that larger molecules are excluded from small pores in the gel packing while the shorter molecular chains permeate the porous media or gel and reside there for various lengths of time. Proper selection of packing with a suitable range of pore sizes results in such residence times in the porous structure that complete separation of the homopolymer takes place as the polymer molecules pass down the column.

Gel-permeation chromatography requires (a) a means of forcing solvent through packed columns, (b) a means for injecting the sample, (c) columns packed with porous media, (d) porous media of suitable ranges of porosity for the molecular weights of interest, (e) calibration standards for converting elution volumes to molecular weights, (f) a device for monitoring eluant volume, and (g) a detector for measuring concentrations in the eluate.

Several excellent reviews on GPC have been published. Cazes [11] gives a fine general description. The many aspects of this technique — its special features, theory and mechanism, the gels, experimental technique, evaluation of data, and the like — are discussed very ably by Altgelt and Moore [12, 13], while Determann [14] enters into particulars on some of the fundamentals. A more recent and thorough discussion can be found in a volume edited by Altgelt and Segal [15].

Use of the term "gel-permeation chromatography" as a name for this technique has become widespread. Other, less used, names such as "molecular-sieve chromatography" or "exclusion chromatography" are quite acceptable. The term "gel filtration" is seen extensively. It originated in work by biochemists and others working with natural products in aqueous solutions. It is clearly a misnomer because in filtration the larger particles are retained by the filter medium — and completely so if the porosity of the medium is low enough — while the smaller particles freely pass through the filter medium along with the solvent. (Something very similar is found in osmosis with semipermeable membranes and in dialysis.) Furthermore, "gel filtration" infers that the gel serves as a filter medium, which is not correct. It is unfortunate that the continued use of this misnomer has not been curtailed in view of its inappropriateness.

B. Extent of Application

For the cellulose chemist gel-permeation chromatography has provided the long-desired means of rapidly obtaining DP distributions of cellulose and its derivatives. In comparison with the more laborious and tedious fractionation methods hitherto available, the GPC technique is a breakthrough.

Perhaps the first published results of the successful application of GPC in the area of cellulose chemistry appeared in 1964; water-soluble hemicelluloses from wood pulp were fractionated on Sephadex columns to verify the practicability of the GPC technique [16]. In 1965 Kringstad [17] reported further work with hemicellulose and Sephadex, and Meyerhoff [18] described an attempt to relate elution volume to molecular parameters of cellulose nitrate dissolved in tetrahydrofuran (THF). Meyerhoff seems to have been the first in the field of cellulose chemistry to use the commercial Waters instrument with crosslinked polystyrene (Styragel) for the column packing. The sole paper published in 1966 described an attempt to correlate GPC data obtained with the Waters instrument with DP data obtained by conventional viscometry [19].

Commencing with 1967 there has been a steady stream of publications and of papers presented at meetings. The research that has been described covers a wide spectrum, ranging from theoretical considerations of column calibration on one hand to GPC applications in wood-pulp and textile technology on the other. It is this wide spectrum of GPC literature relating to cellulose and its derivatives that will be explored here. Because of the essential differences that are involved, the gel-permeation chromatography of cellulose itself will be handled separately from that of its derivatives.

IV. GEL-PERMEATION CHROMATOGRAPHY OF CELLULOSE

A. Solvents and Column Packings

Because unsubstituted cellulose is insoluble in water, cellulose solvents compatible with GPC gels had to be found before GPC could be used to fractionate it. One may restate this by saying that there was the problem of finding suitable GPC gels that could be used with cellulose solvents. Cellulose solvents suitable for GPC applications (cuen, cadoxen, and FeTNa) swell Sephadex gels so strongly that fractionation with these cross-linked dextrans is not feasible. On the other hand, the crosslinked poly-styrene gels (Styragel) that are used with THF solutions of polymers and cellulose derivatives swell too little to be useful. Porous glass or silica beads of controlled pore size (e.g., Bio-Glas) are unsuitable because they are highly degraded by the strongly alkaline solvents. It was Valtasaari [20] who found that FeTNa diluted with 0.2 M sodium tartrate minimized the extreme swelling of Sephadex gels (G-25 to G-200) so that fractionation of a sort could be achieved. Dry gel was immersed in the sodium tartrate solution until swelling ceased; FeTNa was then added, and the swollen gel was then packed into the column. Diluted FeTNa (1:1) was used to form the slurry and also as the eluant.

Pettersson and co-workers worked with cadoxen as the preferred solvent. Bio-Gel P-300, a polyacrylamide gel, was the first gel used for the column packing [21]. Dry grains of gel of suitable mesh size were allowed to swell in an excess of cadoxen (diluted 1:1 with distilled water). The column was packed with the gel slurry, after which the diluted cadoxen served as the eluant. However, Simonson [22], who worked with an eluant composed of glycocoll buffer + 1 M NaCl (pH 10.8), found that Bio-Gel P-100

that had been stored for several months in the eluant contained a large
number of carboxyl groups. On the other hand, fresh gel swollen in 1 M
NaCl contained no carboxyl groups. The presence of these charged groups
resulted in a high degree of gel swelling. Although Simonson worked with
Bio-Gels P-10, P-30, P-60, and P-200, he made no mention of their
swelling, and therefore one could assume that these too would be similarly
affected by an alkaline eluant. In a later investigation of theirs, Pettersson
and co-workers [23] confirmed that with cadoxen the void volume (interstitial volume) of the column decreased because of the increased swelling
of the gel. This effect was said to reduce the control of flow rate and to
lead to difficulties in reproducing results.

Pettersson and co-workers also investigated the suitability of agarose
gel (Bio-Gel A-50m and Sepharose 2B) for use with cadoxen [23-26]. As
the gel was supplied in water, the water had to be replaced by cadoxen
solvent (cadoxen-water, 1:1 by volume). Agarose gel is a crosslinked
galactose. It was found that about 4% of the gel dissolved in the solvent
initially (detected by analysis for galactose in the eluate), but that after this
initial loss the gel bed was stable for many weeks [23].

Sephadex gels, utilized for cellulose in only one instance, as already
noted, have been used extensively to fractionate hemicelluloses. Hemicelluloses are water-soluble polysaccharides that are derived from sugars
other than glucose and may have DPs as high as 150. Unlike cellulose,
many hemicelluloses are partially acetylated or contain many carboxyl
groups. Sephadex columns for use with hemicelluloses are prepared in the
usual manner. Addition of sample solution to the column, elution of the
sample, and collection of the fractions are performed in a manner similar
to the corresponding operations for proteins. The sample may be dissolved
in distilled water or in a salt solution, either of which can be used as the
eluting material. The crosslinked gels that have been used successfully
with hemicelluloses are Sephadex G-25, G-75, G-100, and G-150.

B. Application

1. Cellulose

Valtasaari's pioneering work [20] with Sephadex gels and the cellulose
solvent FeTNa was devoted almost entirely to linear dextran, the 1→6
polymer of glucose; very little work was actually done with cellulose,
which is the 1→4 polymer. Linear dextran was selected as the model
compound to substitute for cellulose because (a) the FeTNa solutions of
the dextran had refractive properties similar to those of cellulose solutions,
(b) the dextran had a considerably lower solution viscosity than cellulose,
which allowed work with more concentrated solutions, and (c) detailed
information was available on the fractionation of water solutions of dextran

on Sephadex. Valtasaari used refractometry to provide a measure (in terms of refractometer reading) of the quantity of solute present in the collected fractions and viscometry for estimation of the molecular weight of the fractions. This was done for the dextran, but not for cellulose. She was successful in using this method to fractionate a linear dextran of $\overline{M}_w$ 39,500 ($\overline{DP}_w$ 244) and a cellulose, origin and prehistory unspecified, of $\overline{M}_v$ 45,000 ($\overline{DP}_w$ 278). It was concluded from these experiments, however, that the FeTNa-Sephadex system could be applied only to low-DP celluloses because of the high viscosities of cellulose solutions and that for celluloses of higher DP other gels would have to be sought.

In contrast, Pettersson and co-workers successfully fractionated cellulose of $\overline{DP}_v$ 780, 1180, and 2000 with cadoxen and polyacrylamide gel [21]. These celluloses were bleached sulfite pulps of spruce wood. Upward flow was used with a syphoning head of 7 cm. The concentration of cellulose in the collected effluent fractions was determined by a modification of the colorimetric orcinol test, using a cellulose calibration curve. Plots of the concentration of cellulose versus elution volume for the three pulps showed clearly that the elution volume at the curve maximum (peak volume) decreased with increasing DP of the sample. Curves obtained with 50:50 mixtures of the samples demonstrated the quantitative additivity of the fractionation. Because of the lack of a calibration procedure for converting elution volume to DP, these fractionation curves only indicate the chain-length distributions in the pulp samples. However, a complete fractionation by this procedure was achieved in less than 40 h (at a flow rate of 4 ml/h) using a 0.5% solution of cellulose. A comparable determination by conventional precipitation fractionation requires days of careful manipulations, close control of temperature, and handling of large volumes of liquid; reproducibility of results is poor; low-DP material is lost, and fractions are not as sharp as may be desired.

Regenerated celluloses in the form of viscose rayon-staple fiber and filament, viscose sausage tubing, and cuprammonium rayon of $\overline{DP}_w$ 320 to 900, dissolved in cadoxen, are claimed to have been fractionated by agarose gel, Bio-Gel A-50m [24, 25]. Cellulose concentration in the eluted fractions was determined by the same orcinol colorimetry method mentioned in the preceding paragraph. Bio-Gel A-50m is the only agarose gel that has been extensively utilized so far. It was found that Bio-Gel A-5m was unable to satisfactorily fractionate native cellulose (bleached sulfite spruce pulp) of $\overline{DP}_v$ 780 [23] because of exclusion of so much of the longer chain lengths by the gel. This is evinced by the very steep rise in the chromatogram immediately after the void volume has been exceeded. However, even Bio-Gel A-50m was found to display some exclusion of longer chain lengths of the spruce pulp [23]. (Note: The $\overline{DP}$s of 250 and 500 given in Ref. 24 were obtained by use of improper Mark-Houwink constants, which give $\overline{DP}_n$. The more correct data for the samples, $\overline{DP}_w$ 323 and 765, are given in

Ref. [25]. The terms $\overline{DP}_W$ and $\overline{DP}_V$ are often used interchangeably; theoretically they are not identical, but usually because of the constants used the values are of quite similar magnitude.)

Examination of the chain-length distribution curves for the regenerated celluloses [24, 25] suggests that Bio-Gel A-50m had not done so well either with regenerated celluloses of $\overline{DP}_W$ greater than about 500, although it satisfactorily fractionated the regenerated celluloses of lower DP. Pettersson even stated [24] that the distribution curve for a Bemberg (cuprammonium) rayon of $\overline{DP}_W$ 765 ($\overline{DP}_n$ 500 in the reference) indicated that the high-molecular-weight portion of the sample was unable to penetrate the matrix of the agarose gel. The DP of this rayon is very similar to that of the spruce pulp.

Because of the previously stated limitations of Bio-Gel A-50m with respect to incomplete permeation of the gel matrix by the longer chain lengths of spruce pulp and viscose rayon ($\overline{DP}_V$ 780 and 765, respectively) and because of the stated void volume of 75 ml for the agarose gel, the later work of Almin, Eriksson, and Pettersson [26] with this gel presents an enigma. The authors report that a native cellulose, cotton linters, with a very high DP of 7000 was dissolved in cadoxen and passed through a column of agarose gel (Bio-Gel A-50m). (A DP of 7000 for cotton linters is indeed very high; unfortunately for such an important sample, however, no other information is given.) Almin et al. stated that "the main part of the cotton linters was obviously excluded from the gel matrix and appears with the void volume of the column." Such results are most confusing because the same gel partially excluded celluloses of DP 765 and 780, and hence cellulose of DP 7000 should have been totally excluded! There seems to be something peculiar about the performance of this gel with this cellulose. Along with the published chromatogram for the DP 7000 cellulose there is included in the figure the chromatogram obtained for glucose, the monomer of cellulose, DP 1. The latter chromatographic curve should be very narrow, but is shown as a very broad distribution.

One may well query the amount of work done with cellulose and the agarose gel, Bio-Gel A-50m, which is so unsatisfactory for high-DP cellulose, whereas the polyacrylamide gel, Bio-Gel P-300, shows better performance. When the very same sulfite spruce pulp of $\overline{DP}_V$ 780 was fractionated with the agarose gel [23] and the polyacrylamide gel [21], the agarose chromatogram exhibited very poor fractionation in comparison with the polyacrylamide chromatogram. Eriksson, Pettersson, and Steenberg [23] state that complete fractionation of most cellulose samples was not obtained with the polyacrylamide gel because the largest molecules appeared in the void volume of the column (given by them as about 63 ml). On the other hand, the work of Eriksson, Johanson, and Pettersson [21] with the spruce pulps of $\overline{DP}_V$ 780, 1180, and 2000 is not in agreement with that statement. It appears that, because the agarose gel gave better fractionation of carboxymethylcellulose (CMC) than did the polyacrylamide gel, these results were extrapolated to cellulose, and further studies with poly-acrylamide gel and cellulose were no longer pursued. Even this reason has to be qualified because Ericksson et al. found that complete fraction-ation of CMC by agarose gel could not be realized with CMC of $\overline{DP}_V > 380$ (see next subsection).

2. Carboxymethylcellulose

The Swedish investigators studied the fractionation of the carboxymethyl derivative of cellulose [23, 24, 26]. They stated that one advantage of CMC is that it is a good model substance for cellulose due to structural similarities between CMC and cellulose in the cadoxen solvent [23]. It is not clear why a model compound is needed at all. Carboxymethylcellulose differs from cellulose by having carboxylic acid side groups:

Cellulose:

$$cell-(OH)_3$$

Carboxymethylcellulose:

$$cell-(O-CH_2-COOH)_n, \quad n = 1, \ 2, \ or \ 3$$

The role of the substituent carboxyl groups in influencing the permeation of the polyacrylamide gel by CMC has apparently been overlooked. These groups can cause one gel to be more suitable than the other for the fraction-ation of CMC.

Because Eriksson and co-workers [23] found that low-molecular-weight CMC ($\overline{DP}_V$ 380) appeared in the void volume of the polyacrylamide gel (Bio-Gel P-300), they investigated the agarose gel (Bio-Gel A-50m). They found that complete fractionation with the latter gel, however, could be achieved only for CMC with a $\overline{DP}_V$ not exceeding 380. For CMC of higher DP (e.g., 666 and 815) chromatograms displayed strong evidence that longer chain lengths were appearing in the void volume. Pettersson [24] presented a CMC chromatogram (from agarose gel) that markedly resembled that of a cuprammonium rayon of $\overline{DP}_V$ 765. In both chromatograms exclusion of longer chain lengths by the gel was indicated.

Almin and co-workers [26] carried out some interesting experiments. They fractionated a CMC by gel-permeation chromatography (agarose gel) into seven fractions and used these fractions to establish a calibration curve. In order to regenerate the CMC fractions, the CMC bulk polymer was dissolved in 0.2 M NaCl solution. For an unexplained reason the 2% agarose gel used for this fractionation was Sepharose 2B. The $\overline{DP}_W$ of the fractions, measured by viscometry of their cadoxen solutions, ranged from 71 to 710. Fractions, as well as bulk polymer, were dissolved in cadoxen and passed through the GPC column to obtain the chain-length distributions. In this case the 2% agarose gel used was Bio-Gel A-50m. It is notable that

the total effluent volume for the bulk polymer dissolved in cadoxen and
passed through Bio-Gel A-50m was about 335 ml (flow rate 4 to 5 ml/h-
cm^2); for the bulk polymer dissolved in 0.2 M NaCl solution and fractionated by Sepharose 2B the volume was over 1050 ml (flow rate 3 ml/h-cm^2).
Since both gels were 2% agarose gels and the columns were similar, these
observed differences seem to raise questions concerning agarose gels and
solvents for GPC work with CMC and other cellulose derivatives that are
soluble in aqueous solvents.

The chromatogram of the bulk polymer of $\overline{DP}_W$ 480 displayed partial
exclusion of longer chain lengths, which would be expected on the basis of
the limitation already established ($\overline{DP}_W$ 480 exceeds $\overline{DP}_W$ 380); the same
was observed for fraction 1 with a $\overline{DP}_W$ of 710. However, chromatograms
for fractions 2, 3, and 4 ($\overline{DP}_W$ 520, 600, and 440, respectively) did not
show the expected exclusion, but were symmetrical Gaussian curves! The
chromatogram for fraction 7 was skewed in the direction of the shorter
chain lengths. Except for the bulk polymer (fraction 1) and fraction 7, the
data are not in agreement with those already published, which were previously obtained with agarose gel of the same designation. This observation
suggests that an investigation of the permeability limits of various lots of
Bio-Gel A-50m would be in order.

Almin and co-workers [26] attempted to establish a calibration curve
for converting elution volume to DP so that the GPC elution curves could be
transformed into DP distribution curves and to correct the experimental
chromatograms for zone or band broadening. By plotting log $\overline{DP}_W$ for the
fractions versus the corresponding peak elution volumes, they obtained a
linear relationship over the DP range. Almin [27] also obtained a linear
relationship by plotting $\frac{1}{2}$ log $\overline{DP}_W \cdot \overline{DP}_n$ against elution volumes. The two
plots were not parallel, but were slightly displaced and approached each
other at very low DPs. Closer examination of the data points establishing
the two relationships suggests that a single line could be passed through
both sets of points. Although the authors measured the $[\eta]$ values and
elution volumes for the fractions, they did not plot their data in the manner
of Benoit and co-workers [28], that is, log $[\eta]\overline{DP}_W$ versus peak elution
volume. However, when this plot is constructed, a good linear relationship
results; it is again not parallel to the log $\overline{DP}_W$ plot but crosses it at the
approximate midpoint of the DP range. An opportunity seems to be afforded
here for a further test of Benoit's universal calibration [22] using cellulose
and cellulose derivatives soluble in cadoxen. Of course, the calibration
curve based on log $\overline{DP}_W$ is valid only for CMC.

The zone-broadening correction for the experimental chromatogram,
developed by Almin and co-workers for this GPC system, did not seem to
make any significant changes in the shape or peak location of the most
symmetrical chromatogram used as an example. This result supports the
earlier conclusion of Tung [29], confirmed by Pierce and Armonas [30],

that the zone-broadening correction is only minor for broad-distribution polymers.

3. Hemicellulose

Although hemicellulose is not cellulose, applications of gel-permeation chromatography in its fractionation are included because of the very close association of the two materials. Cellulose is the 1→4 polymer of glucose and is insoluble in water and other solvents. Hemicellulose is a general term for the water-soluble 1→4 polymers of sugars other than glucose; these polymers may be made up of more than one sugar (e.g., gluco-mannosan, which is a polymer of mannose and glucose). Hemicelluloses are usually isolated from wood and the woody portions of plant stems. Some hemicelluloses have branched structures as contrasted with the linear structure of cellulose; some have acetyl side groups; others have carboxylic acid groups. Separation of hemicelluloses is important in studies of lignin-hemicellulose compounds and the composition of wood extractives.

Kringstadt and Ellefsen [16] were the first to verify the practicability of using gel-permeation chromatography for fractionating hemicelluloses according to their molecular sizes. However, in their study of spruce wood with Sephadex gels and aqueous solutions, they did not characterize the particular polysaccharide isolated from the wood but used the technique to establish the relationship between the polysaccharide and the lignin. Kringstadt [17] carried this GPC application further to resolve the poly-saccharide into galactoglucomannan and arabino-4-0-methylglucuronoxylan. Simple elution curves or chromatograms were plotted from elution volumes versus spectrophotometric data as well as amounts of $K_2Cr_2O_7$ consumed by the fractions.

Further studies of the above xylan from Norway spruce by Zinbo and Timell [31] established its molecular-weight distribution by extensive use of gel-permeation chromatography. The xylan was resolved into 12 fractions by repeated GPC fractionation on Sephadex G-75, and each of the fractions was characterized with respect to weight percent, intrinsic viscosity, $\overline{M}_n$ and $\overline{DP}_n$, methoxyl content, ratio of arabinose to xylose residues, and number of branches present per average macromolecule. The plot of elution volumes versus log $\overline{M}_n$ was linear within a restricted range; at the high-molecular-weight end of the curve, however, there was clear evidence that exclusion was occurring.

Fractionation of hemicellulose by gel-permeation chromatography affords several advantages over the fractional precipitation method. The only parameter affecting GPC separation is molecular size. In fractional precipitation a serious drawback is that separation is based, at least in part, on chemical composition in addition to molecular weight. Thus fractionation by precipitation is affected by the uronic acid groups of hardwood xylans and by the L-arabinofuranose side chains of softwood xylans. Furthermore, with gel-permeation chromatography, sample size and solution concentration are low. For example, a distribution can be obtained with 300 mg of xylan dissolved in 15 ml of water; from 30 to 45 fractions of 20 ml each will establish the chromatogram, which usually qualitatively approximates a molecular-weight distribution.

Simson, Cote, and Timell [32] also examined arabinogalactans from several species of larch by means of gel-permeation chromatography (Sephadex G-100 and G-150). Like other physicochemical methods (ultra-centrifugation, electrophoresis, etc.), GPC studies indicated the presence of two components in the hemicellulose, arabinogalactan A and arabino-galactan B; but only by the GPC technique were these two fractions separated and isolated for further characterization. Ettling and Adams [33] also separated the two fractions by means of gel-permeation chroma-tography on Sephadex G-25, G-75, and G-100.

Xylans from birch wood and from esparto grass, as well as a glucomannan from pine, were fractionated with polyacrylamide gel (Bio-Gel P-300) and with agarose gel (Bio-Gel A-50m) [23]. The peaks of the three hemicelluloses were better resolved by the polyacrylamide gel than by the agarose. It might be pointed out that the GPC fractionations were performed with only 3 ml of 0.3% solutions of the hemicelluloses. It is unfortunate that the crosslinked dextran gels (Sephadex) were not included in this series.

V. GEL-PERMEATION CHROMATOGRAPHY OF CELLULOSE DERIVATIVES

A. Derivatives

Carboxymethylcellulose was discussed in Section IV because of its solubility in an aqueous solvent system and because of its fractionation by the hydro-philic gels used for cellulose. Although there are a few other water-soluble cellulose derivatives, none of these has been investigated by GPC methods. Most of the GPC investigations on cellulose derivatives have been concerned mostly with two derivatives that are soluble in organic solvents: cellulose trinitrate and cellulose acetate. However, some work has been done with cellulose tricarbanilate [34] and triesters of the homo-logous series propionate to heptanoate [35].

Cellulose nitrate per se has not been of interest; GPC separations have been for the purpose of studying cellulose that had been nitrated rather than the cellulose nitrate itself. Usually, when cellulose is esterified or etherified, the polymer undergoes a decrease in DP. However, a cellulose derivative that is soluble in organic solvents and undergoes little loss of DP during esterification can be obtained through nitration according to the procedure of Alexander and Mitchell [36]. Such a cellulose derivative has been invaluable in earlier studies of cellulose by viscometry, osmometry, ultracentrifugation, and precipitation fractionation. In GPC studies of cellulose where separations are made for determining molecular-weight distributions and average DPs, fractionation is actually performed with the cellulose trinitrate ester in solution; results are translated to the cellulose molecule and are so referred to. Thus we find under the heading "Gel-Permeation Chromatography of Cellulose Derivatives" discussions of the gel-permeation chromatography and the molecular-weight distributions of cellulose.

B. Methodology

1. Instrumentation

Cellulose derivatives are soluble in a variety of organic solvents, but the principal solvent used in GPC applications to studies of cellulose is THF. The reasons for this are (a) THF is the solvent having the best compatibility [11, 13] with the crosslinked polystyrene gel (Styragel) and (b) the cellulose derivatives are readily soluble in THF. Other solvents for cellulose nitrate and acetate, such as acetone or ethyl acetate, are incompatible with the packing.

Instrumentation for gel-permeation chromatography is commercially available from several manufacturers, and a compact unit with several options can be obtained from Waters Associates*. Altgelt and Moore [13] have provided general information on GPC instrumentation, and Cazes [11] has described the analytical-scale and analytical-preparative-scale gel-permeation chromatographs in detail. With the analytical-scale instrument a complete chromatogram for a sample can be obtained in about 3 h using only 2 ml of a 1/8 to 1/4% solution of the polymer.

Accumulation of a large amount of fractionated material is a lengthy process because of the very low polymer concentration in eluted fractions, even in the peak fraction. This time can be shortened by using an analytical-preparative-scale instrument.

Concentration of sample components in the eluting solution emerging from the columns is monitored with a differential refractometer. Because

---

*Use of a company or product name does not imply approval or recommendation of the product to the exclusion of others that also may be suitable.

the difference between the refractive index of THF and that of even the most concentrated (peak) cellulose nitrate fraction is quite small, the refracto-meter must be operated at its highest sensitivity, optical and electronic, to obtain measurable outputs as the sample is fractionated. Use of maximum settings is a disadvantage because of accompanying base-line shifts. Differ-ences in the refractive indices of methyl ethyl ketone (MEK) and MEK solutions of celluloses trinitrate are much larger than those with THF solutions [37], and hence with MEK solutions lower sensitivities may be used for the same output. Despite the danger of channeling that might accompany replacement of THF by MEK, the advantage of reduced base-line shifts suggests the need for further studies with MEK.

The recorded chromatogram is not a molecular-weight or chain-length distribution, although it closely resembles them. However, it is a useful qualitative indication of the molecular distribution in the polymer. For quantitative results the count must be correlated with molecular weight or molecular size (average DPs), and the refractive-index difference must be converted to weight of polymer in the fractions. For cellulose, the conversion of count to DP has been the subject of much study.

Schurz and co-workers [38, 39] built a laboratory apparatus using glass columns, filled the columns with pairs of Styragels of different permeability limits, and varied the ratio of one Styragel to the other. The apparatus closely resembled the columns used normally with Sephadex gels, except that they were jacketed for temperature control and a syphon was used as a fraction cutter. The sample was placed at the top of the column, and the solvent was forced through the column by nitrogen pressure. The fractions were taken off and analyzed by differential refractometry as well as by the orcinol colorimetric procedure to determine polymer concen-tration. Measurements of intrinsic viscosity were also made on the fractions. The chain-length distributions of the samples under study were developed by plotting the intrinsic viscosities of the fractions versus the cellulose concentrations.

2. Calibration Standards

a. Cellulose Standards. The GPC technique is not an absolute technique, and as with viscometry, calibration is required against materials of known molecular weight or DP before experimental measurements can be evaluated quantitatively. Calibration standards for gel-permeation chroma-tography must be well-characterized polymer fractions of narrow molecular-size distributions. Standards prepared from cellulose or cellulose nitrate would permit direct conversion of count to DP, but unfortunately these are not commercially available. Suitable cellulose fractions are difficult to obtain, and there are problems relating to characterization and breadth of distribution of the fractions [40]. Yet, as a first approximation, standards of this type are useful. Meyerhoff [18, 41, 42] used seven cellulose nitrate fractions of undefined prehistory in his study of molecular parameters and gel-permeation chromatography. Muller and Alexander [43] were too vague regarding their cellulose calibration standards for their results to be meaningful in terms of DP. However, in later work [44, 45] they established a GPC calibration curve based on 16 cellulose nitrate fractions prepared from cotton cellulose and separated by fractional precipitation, and on the nitrates of cellobiose and cellopentose. These fractions were characterized by viscometry in ethyl acetate. The elution curves of the fractions appear quite broad, but the curvilinear plot of peak count versus log $\overline{DP}_V$ affords direct conversion of count to DP over the DP range 2 to 4000. Even though there is still much to be desired in these standards, they are perhaps the best in the way of cellulose standards that have appeared so far.

Linear plots of peak count versus log $\overline{DP}_V$ over shorter ranges of DP were obtained by Huang and Jenkins [46] (see also Jenkins [47]) using cellulose nitrate fractions from two types of dissolving wood pulps. Precipitation fractionation with acetone-water gave 11 fractions in the DP range 165 to 1660 for one nitrated wood pulp. Six samples of the other wood pulp were submitted to γ-ray irradiation to produce celluloses of DP 60 to 1300; then these were nitrated. The fractions separated by precipitation were characterized viscometrically in acetone, while the $\overline{DP}_V$ of the irradiated celluloses was measured in cuen. Polymolecularity ratios for all fractions ranged from 2 to 4, indicating broad distributions that were at least as broad as those of the initial wood pulps.

A more pragmatic approach for cellulose calibration standards was taken by Phifer and Dyer [48]. Instead of fractionating a nitrated cellulose and characterizing the fractions, they selected and nitrated a series of eight celluloses in the DP range 50 to 1500. These included pulps, yarns, and an Avicel microcrystalline cellulose. The DPs of these celluloses were obtained viscometrically, and the peak counts of the chromatograms were plotted against log $\overline{DP}_V$. In the curvilinear relationship obtained the curve is concave upward, rising slightly in the low-DP region. This curve is opposite in character to that obtained by Alexander and Muller [44], whose curve is concave downward, dropping rapidly in the low-DP region.

Direct calibration in the very low range of DP was carried out by Chang and co-workers [49] with nitrated samples of glucose (DP 1), cellobiose (DP 2), cellohexose (DP 6), and three hydrolyzed celluloses from tire-cord yarn (DP 13), Fortisan (DP 28), and cotton linters (DP 120).

b. Noncellulose Standards. Whereas cellulose standards are not available commercially, well-characterized polystyrene fractions with very narrow distributions are available. These standards range in $\overline{M}_W$ from 5000 to 2,145,000. A polystyrene standard of 3,500,000 molecular weight,

essential for GPC work with high-DP celluloses, was prepared by the Dow
Chemical Company for H. Mark of Polytechnic Institute of Brooklyn. A
small quantity of this polymer, S-115, was made available by Dr. Mark for
the author's work on high-DP cellulose. For molecular weight in a range
below that covered by the polystyrene standards there are the polyglycol
standards. The polystyrene and polyglycol standards are inexpensive and
readily available. Polystyrene standards have been used extensively by
Segal and co-workers [19, 50-54], Brewer and co-workers [35], Tanghe
and co-workers [55], and Chang [56]. Polystyrene and polyglycol standards
were used by Muller and Alexander [43].

3. Conversion of Elution Volume to Degree of Polymerization

The direct conversion of elution volume to DP when cellulose standards
have been used follows simply from the well-established relationship
between count and log M observed within a polymer series. Should well-
characterized cellulose standards become readily available, this method
will become the method of choice.

The use of noncellulose standards to obtain DP data for cellulose
samples introduces problems. The early procedure for converting elution
volume to $\overline{M}_w$ and $\overline{M}_n$ is described clearly by Cazes [11] and by Harmon
[57]. This procedure involves taking the length of the extended molecular
chain and a conversion factor (the "Q-factor") to convert chain length in
angstroms to molecular weight and is satisfactory when calibration standards
and the sample under investigation are the same polymer (e.g., polystyrene).
If standards and sample are not the same polymer but have similar hydro-
dynamic properties in solution, this procedure will still give acceptable
results. However, if the calibration standards and the sample are unrelated
polymers (e.g., polystyrene and cellulose nitrate) having different hydro-
dynamic properties in solution, this procedure leads to very erroneous
data. In Segal's early work [19, 50, 51] very high DPs, ranging from
three to four times greater than the $\overline{DP}_v$ values obtained in cuen and
cadoxen, were reported for various celluloses. He considered these results
acceptable at the time because cellulose data of similar magnitude had been
found in the literature; the earlier data were considered to be confirmed
by the GPC data. Meyerhoff, however, cautioned against use of the polystyrene standards for cellulose [42] and believed that the high DPs obtained
by Segal arose from use of those standards. A similar conclusion was
reached by Huang and Jenkins [46]. Later work by Segal, Timpa, and
Wadsworth [52] proved that the high DPs were indeed erroneous and that
the error arose from using the procedure of extended molecular chain
lengths and Q-factor for converting count to DP. Chang [56] attempted to
retain this conversion procedure by applying a highly empirical "$\overline{Q}$-factor"
that he obtained from viscometric and osmometric measurements on
oxycellulose prepared by treating purified cotton linters with sodium

hypochlorite. His presentation and application of the factor are too ill-based for his proposal to be of much value.

The proposal of Benoit and co-workers [28], on the other hand, was a breakthrough for GPC studies of cellulose. Benoit et al. established a relationship that is valid for any polymer molecule, regardless of its chemical nature and structure. This relationship is the plot of elution volume versus log $[\eta]M$ (the product $[\eta]M$ is designated coil size; $[\eta]$ is intrinsic viscosity and M is molecular weight). With Benoit's approach, polystyrene standards can be used satisfactorily to calibrate a gel-permeation chromatograph for cellulose studies.

Segal and co-workers [52] applied Benoit's concept in their development of a new GPC technique for cellulose. They utilized a little-known but sound mathematical procedure for obtaining $\overline{DP}_w$ and $\overline{DP}_n$ from the GPC measurements, a procedure entirely different from that used with the Q-factor and extended chain length. Their technique required measurement of the intrinsic viscosities of the more concentrated fractions while at the same time obtaining fraction concentrations from measurements of area on the chromatogram. Viscosity measurements, however, constituted a potential source of error because of the low concentrations of polymer in the fractions, but by making use of this technique these workers refined their earlier data and obtained DP data much more consistent with viscosity data.

Wadsworth, Segal, and Timpa [54] proposed a means of eliminating viscosity measurements on the fractions, thereby removing this source of error and leading the way to fully automated data acquisition. In this work the Benoit GPC relationship was combined with the Mark-Houwink viscosity relationship to give the equation

$$M = \left(\frac{\text{coil size}}{K}\right)^{1/1+\alpha}$$

where M is molecular weight and K and α are the coefficient and exponent of the Mark-Houwink equation for cellulose nitrate in THF. The further refinement of Segal's data that was realized from this development indicated that it has very real potential. A related proposal was made by Chang [56], but the proposal was not developed into a working concept.

VI. ADVANTAGES OF, AND LIMITATIONS IMPOSED BY, THE CALIBRATION STANDARDS

A. Cellulose Standards

The advantage of using cellulose standards to calibrate the instrument has been noted in the preceding section. However, there are also limitations in their use. It is doubtful that fractional precipitation or fractional dissolution of cellulose or cellulose derivatives can yield the truly narrow-distribution material that is required, and such techniques as acid hydrolysis or any other form of degradation yield only broad-distribution material. This limitation, however, may not be as bad as it seems because there are calibration procedures based on the use of broad-distribution polymers [27, 58, 59]. The preparative-scale gel-permeation chromatograph offers a possible means of obtaining fractions of cellulose derivatives with narrow distributions, but such fractions would be costly because of the large number of fractionations needed and the pooling of fractions necessary to obtain practical quantities of material.

A more serious limitation is the difficulty in properly characterizing the fraction. Usually viscometry is used for characterization, and there are many values for the constants K and α in the literature. Many authors seem to ignore factors important in obtaining accurate DP values (e.g., corrections for shear gradient, nitrogen content, and kinetic energy); this causes discrepancies and inability by others to duplicate results. For example, Alexander and Muller [44] use the equation DP = 75[η] to cover the entire DP range from 2 to 3500 for cellulose nitrate dissolved in ethyl acetate. No other experimental conditions for viscometry are given, and it must be assumed that the conditions of Alexander and Mitchell [36] were employed for the measurement of [η]. Dyer and Phifer [48, 60] used the same equation for obtaining DP. On the other hand, Huang and Jenkins [46] determined DP over the range 60 to 1660 using acetone solutions and the equations log $\overline{DP}_w$ = 1.32 log[η] - 0.86 when DP $\geq$ 1000, $\overline{DP}_w$ = 1.22[η] when DP $\leq$ 1000; the [η] values (given in milliliters per gram) were determined with Cannon-Ubbelohde dilution viscometers. This viscometer does not require a kinetic energy correction, a claim also made by Alexander and Mitchell for their "Cannon-Fenske type" viscometer. Neither Huang and Jenkins nor Alexander and Mitchell made any correction for shear gradient, a critical factor in celluloses of higher DP.

Meyerhoff [18, 41, 42] speaks of [η], DP, and molecular weight but gives no indication of the viscometric procedures used or of the means of obtaining molecular weights for the data shown in his figures.

B. Noncellulose Standards

With the well-characterized, narrow-distribution polystyrene standards, problems encountered with cellulose standards are no longer present. Instrument calibration from laboratory to laboratory is reproducible. However, in order for these standards to be applicable to cellulose, calibration must be carried out according to Benoit's universal calibration concept [28]. This application necessitates measurement of the intrinsic viscosities of the polystyrene solutions that are injected into the instrument. With viscosities and elution volumes determined and having the known molecular weights of the standards, one then plots the calibration curve.

The use of noncellulosic standards for instrument calibration and the procedure for conversion of elution volumes to DP [54] seems to have freed the GPC determination of average DP of cellulose from the uncertainties attendant with viscometric measurements. However, in Wadsworth's equation $M = (\text{coil size}/K)^{1/1+\alpha}$, K and α are the constants of the Mark-Houwink equation $[\eta] = KM^{\alpha}$ for cellulose nitrate in THF, and these were established by using M values that had been determined viscometrically using ethyl acetate solutions of the same cellulose nitrates that were examined in THF.

The application of Benoit's universal calibration to cellulose nitrate and the polystyrene standards has been questioned by Meyerhoff [18, 41, 42], who claims that data for cellulose nitrate do not fall on the polystyrene calibration curve. Study of some of Meyerhoff's data, however, does not verify this.

When molecular weights, intrinsic viscosities, and elution volumes taken from his published figures [18] for polystyrene, polymethyl methacrylate, and cellulose nitrate were replotted as $\log[\eta]M$ versus count, the three polymers fell on three separate lines [37]. However, data for polystyrene and polymethyl methacrylate should have fallen on a single line. The behaviors of the styrene and the methacrylate polymers have been too well verified for this anomalous behavior to be accepted as valid. Meyerhoff has recognized this discrepancy, but makes no further comment on it [41]. Perhaps errors in the molecular-weight determinations or in the viscosity measurements can account for the anomalies in the GPC data. Unfortunately cellulose nitrate has not been given direct attention by Benoit and others studying GPC universal calibration.

VII. APPLICATIONS IN STUDIES OF MOLECULAR-WEIGHT OR CHAIN-LENGTH DISTRIBUTIONS

In polymer chemistry the physical properties of a polymer are known to be related to the molecular chain length. It is known, too, that the average DPs of polymers arrived at by viscometry, ultracentrifugation, or osmometry are misleading criteria of chain length for the purpose of relating chain length to physical properties. Thus two polymer samples can have the same $\overline{DP}_V$ and yet different breaking strengths, flex lives, toughness, or response to chemical treatments. These parameters are functions of <u>DP distribution</u>, not of average DP. For example, the polymer having the greater proportion of long-chain material in the polymer will have greater toughness, higher breaking strength. As the proportions of short- and long-chain material in

the polymer are readily observable in a distribution curve, examination by
gel-permeation chromatography affords a means of relating chain length
to the physical properties of the polymer.

Examination by gel-permeation chromatography of the results of the
chemical treatments of cotton to convert its cellulose I crystallographic
form to the polymorphic forms celluloses II, III, and IV was the first GPC
application to studies of the DP distribution of celluloses. Segal [19]
clearly demonstrated with differential and integral distribution curves the
extent of degradation resulting from treatments with sodium hydroxide, with
ethylamine, and with ethylenediamine-dimethylformamide. This was
followed by another report [61] showing the distributions found in purified
cotton yarn, purified acetylation-grade cotton linters, acetylation-grade
sulfite pulp, Kraft high-α pulp, and several cellulose acetates obtained from
Eastman Organic Chemicals. Of the celluloses, the linters and the pulps
showed broadened distributions that were shifted toward lower DP values,
as a result of the chemical treatments applied. The distributions of the
Eastman cellulose acetates shifted according to their listed viscosity values,
and displayed, peculiarly, a low-level peak in the high-DP region. This low-
level peak was not found in other cellulose acetates that were also examined.
Based on the history of the acetates, the initial cellulose used in their
manufacture was declared to be wood pulp, and the anomalous appearance
of the peak was explained as being dependent on the nature of the cellulose
being acetylated as no such peak appeared for acetates prepared from
cotton linters [62]. Similar evidence of high-DP material in wood pulps was
also found with nitrated wood pulp [50]. In a detailed study of the high-DP
peak Tanghe, Rebel, and Brewer [55] determined that the peak arose from
hemicelluloses in the wood pulp, primarily mannose and xylose. Alexander
and Muller [44] also concluded that hemicelluloses in wood pulp, which
are absent from cotton linters, are responsible for the peak. However,
Tanghe and co-workers showed that linters acetate would display this peak
if the acetylation time with linters was unduly prolonged (over 150 min).
Through GPC study, these workers were able to monitor the reduction of
the amount of this material in wood-pulp acetates and to thereby obtain
products with improved properties.

Gel-permeation chromatography has been invaluable in studies of
wood pulps and linters pulps that are the basic materials for the manufacture
of paper, textile rayon, tire-cord rayon, cellophane, and various cellulosic
plastics. Although Segal reported on the DP distributions of several wood
pulps [50] and Segal and Timpa [51] discussed the gel-permeation chromatography of several wood pulps and their α fractions, these were cursory
studies. On the other hand, the work of Alexander and Muller [43-45] in
this area is quite extensive. Their work clearly differentiates the many
kinds of wood pulps and regenerated celluloses on the basis of their DP
distribution curves and provides information not readily apparent from other

analytical data. Alexander and Muller used the GPC technique to examine the effects of bleaching on the DP distributions of various pulps and to monitor the analysis for α-cellulose. α-Cellulose is insoluble in cold 18% sodium hydroxide solutions, β-cellulose is soluble but recoverable by warming or diluting, and γ-cellulose is soluble but nonrecoverable by warming or diluting. The DP is lowest for γ-cellulose and highest for α-cellulose. The chromatograms very strikingly showed the removal of lower DP cellulose by the alkali extraction. Such applications have practical ramifications. For example, excessive low-DP components in pulp intended for manufacturing tire-cord rayon is highly detrimental to the properties and physical behavior of the product.

Another extensive study by Dyer and Phifer was concerned with the application of gel-permeation chromatography to the viscose process [48, 60, 63]. They showed very elegantly that commercial pulp sheets that should be homogeneous throughout by virtue of the method of manufacture vary significantly in DP distribution along both axes of the sheet. No indication of this nonuniformity was given by the average DPs obtained viscometrically on bulk samples of the same materials. By GPC, variations in DP distribution were also found in different pulp lots from the same manufacturer. These data are important in blending pulps for further processing. If only the bulk DP (i.e., $\overline{DP}_v$) is used as the criterion for uniformity, improper proportions of components will result that can adversely affect the properties of the final product. Dyer and Phifer followed the DP distribution during steeping and alkali-crumb aging. They found that alkaline hydrolysis was relatively unimportant in changing the DP distribution in the steeping step; the major factor was oxidative degradation. In aging of the alkali crumb, alkaline hydrolysis exerted a greater effect.

The study of Dyer and Phifer using gel-permeation chromatography to follow the effects of various treatments on rayon print cloth [60] is another outstanding application of this technique to practical matters. Fabrics made from regular textile and high-modulus rayons were subjected to the following: cold 25% sulfuric acid, cold 8% sodium hydroxide solution, alkaline peroxide bleach, 25 cycles of washing and drying, ultraviolet radiation, heat, and two types of abrasion. Average DPs were calculated from distribution curves, and measurements were made of the physical properties (tenacity, elongation, etc.) of the treated fabrics. Bleaching drastically reduced DP and tensile properties. Mercerization (8% sodium hydroxide) removed some of the shorter-DP cellulose. Severe degradation resulted from abrasion and flexing; this was assumed to be indicative of the susceptibility of longer chains to degradation by mechanical action. Interestingly, the detritus from the abrasion tests showed relatively narrow DP distributions. The finding that heat shifted the DP-distribution curve so that molecular crosslinking was suggested is highly pertinent to textile processes where the curing step involves high temperatures.

Cotton fabrics crosslinked with formaldehyde, as in the durable-press process, have been studied by gel-permeation chromatography to follow molecular changes; these results have been related to textile properties. The shift in DP distributions to lower DP observed for the Form W' (room temperature, high formaldehyde concentrations, 15% HCl catalyst) was attributed to acid hydrolysis brought about by the catalyst [53]. In the Form C process, where heat (curing temperature 160°C) and a latent or Lewis-acid catalyst (e.g., zinc nitrate) were used in the processing [64], the expected shift of the DP distribution toward lower DP occurred and was accompanied by broadening of the distribution with increasing formaldehyde and catalyst concentrations in the treating bath. With higher formaldehyde contents, more intermolecular linkages resistant to removal were found in the treated cotton. Chromatograms showing evidence that crosslinking or molecular chain extension had been induced by heating water-wet cotton fabric at the curing temperature confirmed the findings of Dyer and Phifer on heat-induced crosslinking.

Brewer and co-workers [34] used gel-permeation chromatography in conjunction with precipitation fractionation to examine cellulose acetates having the same intrinsic viscosities but prepared by two different methods. In one case cotton linters were degraded by acid hydrolysis prior to acetylation; in the second case the linters were acetylated, and then degradation was permitted to take place in the acetylating medium. To ensure that the fractionation would be by DP and not by degree of substitution (DS), the acetates were converted to the tricarbanilate for both GPC and precipitation fractionation. The GPC differential curves for the two preparations were identical and in close agreement with the distribution curves obtained by precipitation fractionation. Unfortunately the theoretical implications of these results were not explored.

The GPC parameters for cellulose nitrate samples containing 13.55 to 13.81% nitrogen (14.14% nitrogen corresponds to complete nitration, DS 3) were shown by Segal and co-workers [65] to be hardly affected by this variation in substitution. The effects of DS and of primary hydroxyl content of cellulose acetate and propionate, as well as the effect of substituent-group size, on the apparent molecular sizes of the esters in GPC fractionation were studied by Brewer and co-workers [35]. As expected, molecular size increased with the DP of the acetate and the propionate. Increasing the acetyl content from 38 to 42% at constant DP had essentially no effect on molecular size, while increasing propionyl content from 38.7 to 51.5% caused only a slight increase. A similar slight increase in size was also observed on increasing the size of the substituent group from propionate to heptanoate. Changing the proportions of the primary hydroxyl group of the acetate from 46 to 19% had no effect on molecular size. Excellent agreement was found between calculated and observed gel-permeation chromatograms obtained with a blend of cellulose tripropionates having high and low intrinsic viscosities and for a blend of hydrolyzed cellulose acetates with high and low viscosities.

Gel-permeation chromatography has served well also in fundamental studies of cellulose. By this means Muggli [66, 67] showed that the cellulose molecules in ramie lie in the extended-chain, rather than in a folded-chain, conformation. One thousand cross sections of 2-μ thickness were cut from ramie fiber and converted to the tricarbanilate for GPC study. The molecular-weight distribution determined by the GPC method remained Gaussian and shifted toward shorter chain lengths, confirming results that were predicted from dimensional considerations of the specimens and from the conformation of the cellulose molecule.

Hydrolytic degradation of cotton cellulose by acid was shown by Segal [50] to cause a marked shift of the DP distribution curve toward low DP and a considerable broadening of the distributions. This behavior was confirmed by Rinaudo and co-workers [68]. Rinaudo also studied the enzymatic hydrolysis of cotton, ramie, a nonstretched rayon, and Whatman powdered cellulose. The DP distributions of the rayon and of the powdered cellulose were unchanged after 48 h of enzymatic attack, but the samples showed considerable losses in weight. The DP distributions of ramie and of native cotton were only slightly affected by the enzyme. However, the same cotton kept wet after swelling with 12% sodium hydroxide solution behaved differently: there was a marked changed in the DP distribution curve and a large weight loss for the sample. The changes effected by enzymatic attack were examined by gel-permeation chromatography after 1, 10, 24, and 48 h of exposure to the enzyme. The DP distribution progressively shifted toward low DPs and broadened, while at the same time the peak at the high-DP end of the chromatogram became less and less apparent. These data were used in proposing a mechanism for enzymatic attack on high-molecular-weight cellulose.

Gel-permeation chromatography has an important role now in cellulose research. With this technique one can obtain molecular-weight data more rapidly and more accurately than was hitherto possible. The technique is unequaled for qualitative comparisons of DP distributions, while quantitative data will become more accurate and precise as calibration problems become resolved. The need for an organo-soluble derivative for handling high-DP celluloses, however, retards greater utilization of the technique because of the extra operations involved. Gel-permeation chromatography has proved its worth in the industrial field; it is even more valuable in fundamental research. In the latter area many problems remain. For example, the GPC technique can furnish answers to questions regarding the DP of cotton cellulose as the fibers develop in the boll and as the fibers dry with boll opening. Still unresolved is the matter of strength variation that is found among cotton varieties which may arise from growing conditions. These are fertile areas for research with gel-permeation chromatography in the hands of the cellulose chemist.

REFERENCES

1. D. M. Cates and T. H. Guion, in Analytical Methods for a Textile Laboratory, 2nd ed., J. W. Weaver, ed., American Association of Textile Chemists and Colorists, Research Triangle Park, N.C., 1968, p. 239.

2. H. Mark and G. Saito, Monatsh., 68, 237 (1936).

3. G. R. Levi and A. Giera, Gazz. Chim. Ital., 67, 719 (1937).

4. S. Claesson, Arkiv Kemi, Mineral. Geol., 26A, No. 24 (1949); Discussions Faraday Soc., No. 7, 321 (1949).

5. M. C. Brooks and R. M. Badger, J. Am. Chem. Soc., 72, 1705 (1950).

6. M. C. Brooks and R. M. Badger, J. Am. Chem. Soc., 72, 4384 (1950).

7. R. S. Porter and J. F. Johnson, in Polymer Fractionation, M. J. R. Cantow, ed., Academic Press, New York, 1967, p. 95.

8. P. K. Chatterjee and R. F. Schwenker, in Instrumental Analysis of Cotton Cellulose and Modified Cotton Cellulose, R. T. O'Connor, ed., Dekker, New York, 1972, p. 289.

9. C. A. Baker and R. J. P. Williams, J. Chem. Soc., 2352 (1956).

10. H. Heyemann, Eastman Kodak Co., personal communication, May 1965.

11. Jack Cazes, J. Chem. Ed., 43, A567, A625 (1966).

12. K. H. Altgelt, in Advances in Chromatography, Vol. 7, J. C. Giddings and R. A. Keller, eds., Dekker, New York, 1968, p.3.

13. K. H. Altgelt and J. C. Moore, in Polymer Fractionation, M. J. R. Cantow, ed., Academic Press, New York, 1967, p. 123.

14. H. Determann, in Advances in Chromatography, Vol. 8, J. C. Giddings and R. A. Keller, eds., Dekker, New York, 1969, p. 3.

15. K. M. Altgelt and L. Segal, eds., Gel Permeation Chromatography, Dekker, New York, 1971.

16. K. Kringstad and Ø. Ellefsen, Das Papier, 18, 583 (1964).

17. K. Kringstad, Acta Chem. Scand., 19, 1493 (1965).

18. G. Meyerhoff, Makromol. Chem., 89, 282 (1965).

19. L. Segal, J. Polymer Sci. B, 4, 1011 (1966).

20. L. Valtasaari, Paperi ja Puu, 49, 517 (1967).

21. K.-E. Eriksson, F. Johanson, and B. A. Pettersson, Svensk Papperstidn., 70, 610 (1967).

22. R. Simonson, Svensk Papperstidn., 70, 711 (1967).

23. K.-E. Eriksson, B. A. Pettersson, and B. Steenberg, Svensk Papperstidn., 71, 695 (1968).

24. B. A. Pettersson, Svensk Papperstidn., 72, 14 (1969).

25. B. A. Pettersson and E. Treiber, Das Papier, 23, 139 (1969).

26. K. E. Almin, K. E. Eriksson, and B. A. Pettersson, J. Appl. Polymer Sci., 16, 2583 (1972).

27. K. E. Almin, Polymer Preprints, Am. Chem. Soc. Div. Polymer Chem., 9, (1), 727 (1968).

28. H. Benoit, Z. Grubisic, P. Rempp, D. Decker, and J. G. Ziliox, J. Chim. Phys., 63, 1507 (1966); Z. Grubisic, P. Rempp, and H. Benoit, J. Polymer Sci. B, 5, 753 (1967); Z. Grubisic, L. Reibel, and G. Spach, C. R. Acad. Sci. (Paris), C246, 1690 (1967).

29. L. H. Tung, J. Appl. Polymer Sci., 10, 1271 (1966); Characterization of Macromolecular Structure, Publ. 1573, National Academy of Sciences, Washington, D.C., 1968, p. 268; see also Reference 15, p. 153.

30. P. E. Pierce and J. E. Armonas, J. Polymer Sci. C, No. 21, p. 27 (1968).

31. M. Zinbo and T. E. Timell, Svensk Papperstidn., 70, 695 (1967).

32. B. W. Simson, W. A. Cote, Jr., and T. E. Timell, Svensk Papperstidn., 71, 699 (1968).

33. B. V. Ettling and M. F. Adams, Tappi, 51, 116 (1968).

34. R. J. Brewer, L. J. Tanghe, S. Bailey, and J. T. Burr, J. Polymer Sci. A-1, 6, 1697 (1968).

35. R. J. Brewer, L. J. Tanghe, and S. Bailey, J. Polymer Sci. A-1, 7, 1635 (1969).

36. W. J. Alexander and R. L. Mitchell, Anal. Chem., 21, 1497 (1949).

37. L. Segal, manuscript in process.

38. J. Schurz and J. Haas, Cellulose Chem. Technol., 4, 633 (1970).

39. J. Schurz, J. Haas, and H. Krassig, Cellulose Chem. Technol., 5, 269 (1971).

40. T. E. Timell and E. C. Jahn, Svensk Papperstidn., 54, 831 (1951).

41. G. Meyerhoff, Makromol. Chem., 134, 129 (1970); J. Chromatogr. Sci., 9, 596 (1971).

42. G. Meyerhoff and S. Jovanovic, J. Polymer Sci. B, 5, 495 (1967).

43. T. E. Muller and W. J. Alexander, J. Polymer Sci. C, No. 21, 283 (1968).

44. W. J. Alexander and T. E. Muller, cited in Ref. 15, pp. 429-453.

45. W. J. Alexander and T. E. Muller, J. Polymer Sci. C, No. 36, 87 (1971).

46. R. Y. M. Huang and R. G. Jenkins, Tappi, 52, 1503 (1969).

47. R. G. Jenkins, master's thesis, University of Waterloo, Waterloo, Canada, 1968.

48. L. H. Phifer and J. Dyer, cited in Ref. 15, pp. 465-480.

49. M. Chang, T. C. Pound, and R. St. John Manley, Preprints, IUPAC Int. Symp. Macromol., 5, 86 (1972).

50. L. Segal, J. Polymer Sci. C, No. 21, 267 (1968).

51. L. Segal and J. D. Timpa, Tappi, 52, 1669 (1969).

52. L. Segal, J. D. Timpa, and J. I. Wadsworth, J. Polymer Sci. A-1, 8, 3577 (1970).

53. L. Segal and J. D. Timpa, Textile Chem. Color., 4, 66 (1972).

54. J. I. Wadsworth, L. Segal, and J. D. Timpa, Polymer Preprints, Am. Chem. Soc. Div. Polymer Chem., 12 (#2), 854 (1971); Advances in Chemistry, #125, American Chemical Society (1973).

55. L. J. Tanghe, W. J. Rebel, and R. J. Brewer, J. Polymer Sci. A-1, 8, 2935 (1970).

56. M. Chang, Tappi, 55, 1253 (1972).

57. D. J. Harmon, J. Polymer Sci. C, No. 8, 243 (1965).

58. F. C. Frank, I. M. Ward, and T. Williams, J. Polymer Sci. A-2, 6, 1357 (1968).

59. M. J. R. Cantow, R. S. Porter, and J. F. Johnson, J. Polymer Sci. A-1, 5, 1391 (1967).

60. J. Dyer and L. H. Phifer, J. Polymer Sci. C, No. 36, 103 (1971).

61. L. Segal, paper presented at the 153rd Meeting, American Chemical Society, Division of Carbohydrate Chemistry, Miami Beach, Fla., April 9-15, 1967.

62. L. J. Tanghe, Eastman Kodak Co., personal communication, June 1967.

63. J. Dyer and L. H. Phifer, cited in Ref. 15, pp. 481-491.

64. L. Segal and J. D. Timpa, Textile Res. J., 43, 468 (1973).

65. L. Segal, J. D. Timpa, and J. I. Wadsworth, J. Polymer Sci. A-1, 8. 25 (1970).

66. R. Muggli, Cellulose Chem. Technol., 2, 549 (1968); doctoral dissertation, Eidgenossichen Technischen Hochschule, Zurich (1968).

67. R. Muggli, H.-G. Elias, and K. Mühlethaler, Makromol. Chem., 121, 290 (1969).

68. M. Rinaudo and J. P. Merle, C. R. (Paris), 268, 593 (1969); Eur. Polymer J., 6, 41 (1970); M. Rinaudo, F. B. Arnoud, and J. P. Merle, J. Polymer Sci. C, No. 28, 197 (1969).

Chapter 3

PRACTICAL METHODS OF HIGH-SPEED

LIQUID CHROMATOGRAPHY

Gary J. Fallick

Waters Associates, Inc.
Milford, Massachusetts

I. INTRODUCTION

A. Background

Modern high-speed liquid chromatography (HSLC) is an extremely powerful method for the separation and analysis of complex chemical mixtures. It is also a very versatile technique, encompassing a number of modes, which may be used singly or in combination to achieve the separation. These modes include adsorption (liquid-solid), partition (liquid-liquid), ion exchange, or selective exclusion according to molecular size (gel permeation).

One of the advantages of this wide-ranging method is simplicity of operation. The one fundamental requirement is that all of the components to be separated be in solution. The scale of operation can be matched directly to the immediate problem, employing a few milligrams of sample for routine analytical purposes or gram quantities for the preparation of highly purified compounds.

All types of chromatography are based on the phenomenon that each component in a mixture ordinarily interacts with its environment differently from all other components under the same conditions. In liquid chromatography (LC) a dilute solution of the sample is passed through a tube or column packed with solid particles, which may or may not be coated with another immiscible liquid. With the proper solvent, operating conditions, and packing, some components in the sample will travel through the column more slowly than others, resulting in the desired separation.

Initially, liquid chromatography was a slow separation technique performed in vertical columns by gravity flow. Recent developments have greatly improved its speed and versatility. The increase in speed has been achieved by pumping the solution through the column at inlet pressures of up to and sometimes exceeding 1000 psi. The gains in versatility have come about through the use of smaller diameter, high-surface-area particles and other unique developments in packing structures and surfaces. Use of these recent innovations is the basis of high-speed liquid chromatography.

The chief consequence of the developments leading to high-speed liquid chromatography is a major gain in resolving power. Just as thin-layer chromatography (TLC) evolved as a means of enhancing the separating capability of the original open-column technique, modern techniques provide even greater gains in resolution. Table I shows the relative resolving power of the various LC methods for the same solvent and adsorbent, using an open column as the basis of comparison.

TABLE I

Relative Resolving Power of Various LC Methods

Method	Relative resolution attainable
Open-column chromatography	1.0
Thin-layer chromatography	1.7
Modern high-speed liquid chromatography	7.0
Recycle[a] liquid chromatography	28.0

[a]A technique for increasing the separating capability of modern LC systems without using additional columns (see Section III for details).

B. Apparatus

1. General Remarks

The apparatus required for high-speed liquid chromatography is relatively straightforward. The basic components of a complete LC system are shown in Fig. 1. While solvent is pumped through the column, a dilute solution of the sample, typically 0.1 to 10%, is introduced into the moving solvent, generally by microliter syringe injection through a self-sealing elastic diaphragm or septum. Larger sample volumes or viscous samples are introduced into the system from a special valve fitted with one or more loops of tubing containing the samples. As the solution flows through the column, additional solvent is pumped behind it. For analytical purposes, narrow-bore columns — a few millimeters in diameter — are used with typical flow rates of 0.5 to 9.9 ml/min.

2. Separation

The components separate into bands while passing through the columns. Plug flow is necessary to avoid remixing the bands and loss of separational efficiency. Furthermore, in a well-designed instrument, "dead volume" — the volume of the tubing between the injector and column, column outlet and detector, and flow channels in the detector — must be minimized. Otherwise diffusional spreading or widening of the bands may occur, causing some interlayer mixing and loss of component resolution.

Gary J. Fallick

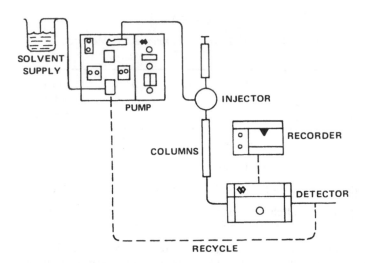

FIG. 1. Liquid-chromatograph flow diagram.

3. Detection

As separated sample components elute from the column, they pass through
one or more detectors. These must be closely coupled to the column in
order to hold down dead volume and avoid band spreading. The most
widely used detectors are based on change in the refractive index or, if
applicable, absorption at a fixed wavelength of ultraviolet radiation. Since
every compound has a refractive index, this property is a universal indicator.
Differential refractometers, such as the Waters R-401, have typical detection
capabilities of 1 ppm.

Absorption at a specific ultraviolet wavelength is a more selective
detection mode for ultraviolet-absorbing compounds. It is very sensitive,
with detection capabilities as low as 0.06 ppm for compounds with high
absorbances. The use of both detector types in series is very effective.
For samples with any amount of complexity, dual detection will almost
always provide more than double the information obtained with either detector
alone. The advantages of tandem detectors are listed as follows [1].

1. Universal detection — 1 ppm

2. Sensitive, selective (254 or 280 nm) ultraviolet detection — 0.06 ppm

3. Detection of unresolved pairs

4. Detection of nonultraviolet absorbers

5. Two-way calibration for increased quantitative precision

4. Recording

The signals from the detectors are recorded as deviations from a base line (Fig. 2). Two-pen recorders are used with dual-detector instruments. The peak position along the curve relative to the starting point denotes the particular component, once its elution characteristics have been determined for the given set of operating conditions. With proper calibration, the height or area of the peak is a measure of the amount of the component present in the sample.

C. Terminology

To perform effective LC separations, a column must have the capacity to retain samples, must separate sample components, and must operate efficiently. Without these characteristics, the separation will not occur within a reasonable operating time or with a practical amount of equipment. Each of these factors has been defined using the recorder trace of detector response versus time (or volume when the flow rate is known).

1. Capacity To Retain Samples

When a sample is injected into the solvent flowing through a column, the inert components that are not adsorbed by the packing will pass through and elute first. The total volume of solvent eluting from the column between the time of injection and the appearance of these unadsorbed species equals the volume of the column not occupied by the packing. This is the void volume V_0 illustrated in Fig. 2.

The elution volume of an adsorbed species is always larger than the void volume. It is usually expressed as volume in excess of the void volume. It is a specific property of the sample component under the given solvent, column, and operating conditions. Components 1 and 2 in Fig. 2 have elution volumes of $V_1 - V_0$ and $V_2 - V_0$, respectively.

The capacity factor of a column system is a measure of sample retention by the column in column volumes. This measure is called k' and is simply the ratio of the component elution volume to the void volume expressed as

$$k'_1 = \frac{V_1 - V_0}{V_0}$$

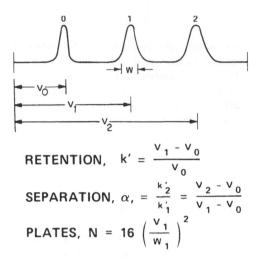

$$\text{RETENTION,} \quad k' = \frac{V_1 - V_0}{V_0}$$

$$\text{SEPARATION,} \quad \alpha, = \frac{k'_2}{k'_1} = \frac{V_2 - V_0}{V_1 - V_0}$$

$$\text{PLATES,} \quad N = 16 \left(\frac{V_1}{W_1}\right)^2$$

FIG. 2. Typical chromatogram.

2. Separation of Sample Components

Since the purpose of liquid chromatography is separation of two or more
components, the relative capacity factors of two components become a
measure of the column's ability to separate them. This is expressed as the
separation factor and is defined as

$$\alpha = \frac{V_2 - V_0}{V_1 - V_0} = \frac{k'_2}{k'_1}$$

If the separation factor is unity, the peaks coincide, and no separation has
occurred. In some modes of modern high-speed liquid chromatography
meaningful resolution is achieved with α values as low as 1.05.

3. Efficiency of Operation

Another key feature of the elution diagram shown in Fig. 2 is the width of
the peaks at their base lines. Wider peaks (band spreading) generally occur
at longer residence times (higher k' values), but other column conditions
can also influence this characteristic.

   For good resolution, narrow base-line widths (minimum band spreading) are certainly desirable, particularly when α is small. An empirical measure of column efficiency is the theoretical plate number N, in which

$$N = 16\left(\frac{V}{W}\right)^2$$

The narrower the peak, the higher N and the more efficient the column. Typical values are 200 to 1000 plates per foot of column for affinity liquid chromatography and 500 to 1500 plates per foot of column for gel-permeation chromatography.

4. Resolution

Resolution R_s is a measure of the number of bandwidths that can fit between the band centers:

$$R_s = \frac{V_2 - V_1}{1/2\,(W_2 + W_1)}$$

Assuming symmetrical (Gaussian) peaks, when $R_s = 1$, peak separation is nearly complete, with only about 2% overlap. This case is shown in Fig. 3.

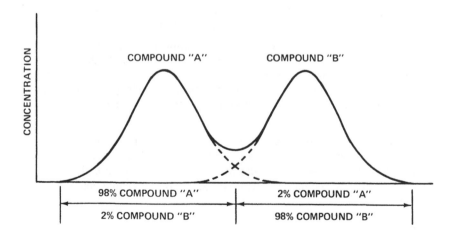

FIG. 3. Symmetrical peaks obtained when resolution $R_s = 1$.

II. DEVELOPING AN LC SEPARATION

A. Defining the Problem

The first step in developing a separation is to have a clear statement of the problem to be solved.

Is it necessary to obtain a quantitative separation of each compound present in a multicomponent mixture, or is there only one component of interest in the sample?

Is it necessary to achieve a general "fingerprint" of the sample for routine screening purposes, or must there be a specific quantitative separation for collection and subsequent analysis by other methods?

Are classes of materials to be separated by size?

Are all the compounds very similar, differing only in isomeric structure?

Is the separation a one-time problem, or will it be adopted as a standard analytical method for routine control work?

Depending on each of these requirements, a satisfactory separation may be achieved after two or three injections. This is especially so when the experimenter is merely seeking a fingerprint or is looking for a single compound, or where the separation will not become an established quality-control method. Alternatively more extensive work is necessary for the total resolution of all components or defining a reproducible control method.

The starting point for selecting a first system is to know the unknown. The more information known about a sample, the better basis there is for beginning. The solubility characteristics of the sample, at least, must be known since it must be in solution to be injected. Knowledge of the best solvents for the sample is a consideration in selecting the first packing to try.

If possible, one should write the molecular structure of all the compounds that are known or suspected to be present. Using infrared spectroscopy, if available, one should determine or verify the functional groups in the sample. Initial screening with thin-layer chromatography should be considered.

B. Selecting the Packing

The choice of packing is relatively simple if initial separation attempts are based on size or ionization differences among the sample components.

1. Size Separation (Gel-Permeation Chromatography)

In dealing with an unknown mixture a useful starting point is initial classification according to molecular size of the various components. Gel-
permeation chromatography (GPC) is a very predictable mode, mechanically
sorting the compounds based on their dimensions in solution. Each size
fraction obtained can then be subsequently analyzed by affinity modes with
less difficulty than might be presented by the original starting mixture.

When the probable molecular dimensions or weights of the sample
components are known, selecting the particular grade of packing is usually
straightforward. For discrete resolution of species in the 100 to 1000
molecular-weight range, differences on the order of 40 to 50 daltons should
exist between components. Column lengths are determined by the magnitude
of the differences. For a packing of given exclusion limit longer columns
are required as the size differences between components diminish. Packings
are selected from Fig. 4 to match the molecular-weight range of the components. The solvent for the sample is the basis for the final choice of GPC
packing. Aqueous systems dictate the use of deactivated Porasil, but all
packings, including deactivated Porasil, are compatible with organic
solvents.

2. Ion Exchange

The use of ion exchange depends on specific knowledge of the ionization
characteristics of the functional groups in the sample. Ion-exchange packings are strong or weak, cation or anion exchangers (Table II).

Ion-exchange packings are available as porous beads or in pellicular
form, consisting of a thin porous layer supported on a solid inactive core.
The pellicular packings have very low exchange capacity, typically 60 μeg/g,
compared with 4 meq/g for the porous type. This dictates the use of very
dilute samples and buffers. In all ion-exchange separations the pH, buffer
composition, and concentration can each have profound effects on the separation. Knowledge of the influence of these factors on the ionization potential
of the sample and on the packing is required. Temperature can also have a
very significant effect on the performance of the packing.

3. Affinity Separations

The greatest uncertainty arises when deciding which affinity packing should
be chosen. In this discussion nonionic affinity packings are considered as a
continuous spectrum of available packing-surface polarities used to achieve
affinity separations. The available packing-surface functionalities (Waters
Associates) are listed in Table III.

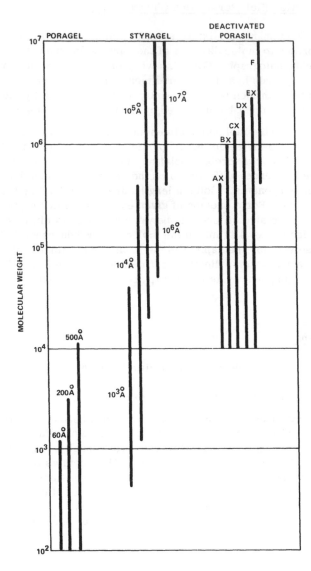

FIG. 4. Effective molecular-weight range of size-separation packings.

TABLE II

Ion-Exchange Packings

Type	Typical packing functional group	Samples	Typical pH range
Anion: Strong	Quaternary ammonium chloride	Free anions from acid or base salts	1-10
Weak	Amine salt	Strong and moderately strong acids, weak bases	4-8
Cation: Strong	Sodiumsulfonic acid	Free cations from acid or base salts	1-10
Weak	Carboxylic acid salt	Strong and moderately strong bases, weak acids	4-8

The procedures to be described are guides to developing a separation and are only starting points. There are always exceptions to almost any rule, and there will be times when a desired separation is achieved with a packing or solvent/packing combination that does not agree with the recommended starting place. Until enough experience is developed with a particular class of compounds to know where to begin, it is wise to begin within the framework described and proceed systematically as indicated.

Consider whether the differences between sample components are primarily steric or primarily in polarity or solubility. If steric, an adsorption packing is probably the best starting point; otherwise a bonded-phase packing may be a better first choice.

If any TLC resolution has been achieved with R_f values from 0.2 to 0.8, it should be a simple matter to transfer the method to high-speed liquid chromatography. Even if the TLC results were marginal or a single spot resulted within the R_f range mentioned, there is every likelihood that the superior resolving power of high-speed liquid chromatography will provide the separation.

TABLE III

Available Packing–Surface Functionalities and Polarities[a]

Packing type	Surface functionality	Polarity
Adsorption		
Porasil (porous)	$-SI-O-Si-OH$	High
Corasil (pellicular)	$-Si-O-Si-OH$	High
Alumina W200-acid (porous)	$-O-Al-Cl$	High
Alumina W200-basic (porous)	$-O-Al-O-Na$	High
Alumina W200-neutral	$-O-Al-O-Al$	High
Bonded Phase		
Durapak OPN	$-C=N$	Intermediate
Poragel PT	$-OH$ and $-C=O$	Low/intermediate
Durapak Carbowax 400	$-(CH_2-OCH_2)$ $n-CH_2OH$	Low/intermediate
Poragel PN	OH and $-C=O$	Low
Poragel PR	Aromatic $-O-$ and $-N-$	Low
Poragel PS	Aromatic $-N-$	Low
Durapak n-octane	$-(CH_2)_7CH_3$	Low
Bondapak:		
phenyl/Corasil	C_6H_5	Very low
C_{18}/Corasil	$-(CH_2)_{17}CH_3$	Very low

[a]These packings are available from Waters Associates, Inc. Porasil, Corasil, Durapak, and Poragel are registered trademarks of Waters Associates, Inc. , and Bondapak is a trademark of this company.

The following considerations apply when transferring TLC results to high-speed liquid chromatography:

1. Use a packing with surface functionality comparable to that of the TLC material, such as Corasil (Waters Associates, Inc.) or Porasil, when a silica place was used

2. If the plate was equilibrated with water vapor in the air, use a low-surface-area packing, such as Corasil I or Porasil D or E

3. If the TLC system was anhydrous, use packings of higher surface area, such as Porasil A or C

4. Begin with a solvent system of lower polarity in the liquid chromatograph than was used in the TLC system

A successful affinity separation is achieved by establishing the proper balance between attraction of the solvent and packing for the sample. In an LC column solvent and packing both compete for the sample. If the sample is more like the solvent than it is like the packing in terms of polarity (or other measures of solvating strength), there will be little retention. When this occurs, it is necessary to make the solvent less like the sample, generally by changing the polarity of the solvent system. An alternative is to change the packing to a type more similar to the sample. Most good separations are achieved by matching the polarities of the sample and packing, and using a solvent that has a markedly different polarity.

There are some instances in which the sample is strongly solubilized, and even a substantial change in solvent polarity does not increase the elution time or improve resolution. In these situations it is often appropriate to use reverse-phase chromatography. In reverse-phase chromatography the sample and packing are very nonpolar, whereas the solvent is relatively polar. This is in contrast to the more commonly used normal-phase liquid chromatography in which a polar packing such as silica is used with nonpolar solvents such as chloroform or isooctane.

The relative polarities of commonly used LC solvents are presented graphically as the eluotropic series. Though polarity is a useful basis for ranking solvents, other characteristics, such as hydrogen-bonding properties and dispersive (London) forces, also contribute to solvating strength. All of these factors are taken into account in the solubility parameter. A discussion of the solubility parameter is beyond the scope of this review, but may be found in most recent physical chemistry texts. Solvents in the eluotropic series with comparable polarity rankings may have substantially different solubility parameters.

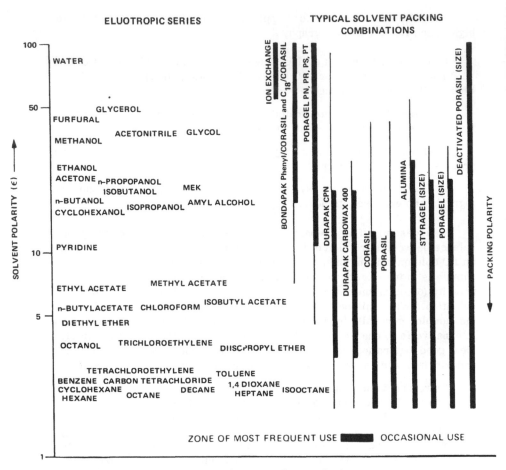

FIG. 5. Eluotropic series and column-packing polarities.

    The packings that are used most often within particular solvent-polarity ranges are shown in relation to the eluotropic series (Fig. 5). Since solubility-parameter differences often exist within a given solvent-polarity range, it is often expedient to screen a series of appropriate solvents with a single packing during the early stages of work on a separation.

    General interactions between sample and solvent as a function of polarity are shown in Fig. 6 and summarized in Table IV. The solvent polarities generally used in the normal phase range from low to medium. Considering a two-component sample mixture (A and B) in the normal phase

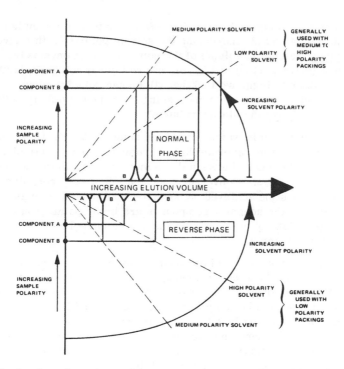

FIG. 6. Sample-solvent interactions in high-speed liquid chromatography.

TABLE IV

Sample-Solvent Interactions as a Function of Polarity

Parameter	Normal phase	Reverse phase
Packing polarity	High	Low
Solvent polarity	Low to medium	Medium to high
Sample elution order	Least polar first	Most polar first
Effect of increasing solvent polarity	Reduces elution time	Increases elution time

(upper portion of the diagram), note that (1) lower polarity sample components (B) elute first and (2) increasing the solvent polarity with a given packing <u>reduces</u> the elution volume of the components. Conversely, in reverse phase (lower portion of Fig. 6), where typical solvent polarities range from medium to high, we find that (1) higher polarity sample components (A) elute first and (2) increasing the solvent polarity with a given packing <u>increases</u> the elution volume.

There are instances in which very polar compounds might also be separated successfully by reverse-phase chromatography, particularly where the major differences between the components are in the nonpolar portions of the molecules. Conversely, separation of very nonpolar compounds might best be achieved on normal-phase polar packings if the differences between the nonpolar compounds are in the polar groups. Obviously, separations of many compounds can be made by either normal- or reverse-phase systems.

Estrogens are an example of compounds that can be separated by both normal- and reverse-phase chromatography. The monoalcohol precursor, the diol, and the triol obey the general model presented. In normal-phase systems estriol is most like the polar silica packing and is therefore held most strongly by it, eluting last (Fig. 7). In reverse-phase (systems) estriol is least like the nonpolar packing relative to estradiol and the monoalcohol. Consequently estriol elutes first in reverse phase, and the elution order of the three compounds is reversed (Fig. 8).

In cases where little is known about a sample, it is still possible to take a consistent approach. Any problem has boundary conditions. In liquid chromatography they are the extremes of packing polarity that are available. The most polar surfaces are found on Porasil and Corasil silica packings. The least polar is $Corasil/C_{18}$.

After the sample has been dissolved and a preliminary size separation performed, one should consider the solvent system that was used. If it has a medium to high polarity, one should begin with reverse phase using $Corasil/C_{18}$. If the preferred solvent is low in polarity, one should start with Corasil II. Figures 9 and 10 show the steps to follow after the first injection and after the chromatogram is obtained.

One note of caution: while the investigator is probing for a workable system with little knowledge of possible sample/packing interactions, some sample components may remain on the column. Consequently, after all apparent elution has occurred from each exploratory injection, a very strong solvent for the sample should be flushed through the column to clean it thoroughly. The next solvent used should then be allowed to equilibrate with the column packing before proceeding with the next injection.

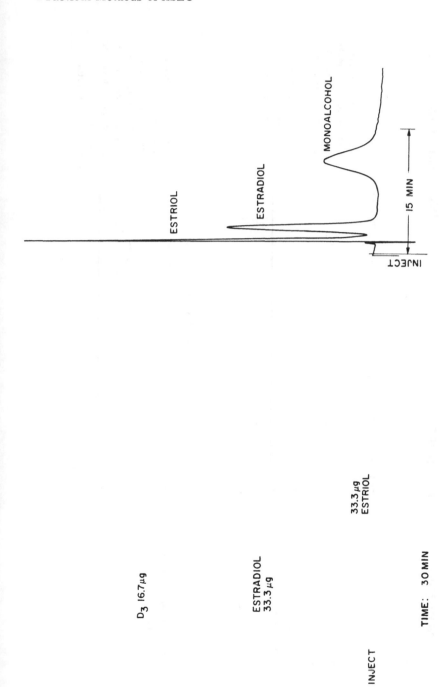

FIG. 7 (left). Estrogen separation by normal phase chromatography. Packing: Porasil A. Solvent: chloroform–acetonitrile (50:50). FIG. 8 (right). Estrogen separation by reverse-phase chromatography. Packing: Bondapak C_{18}/Corasil. Solvent: water–acetonitrile (2:1).

NON-IONIC AFFINITY MODES

ION EXCHANGE

NO RETENTION   ALL COMPONENTS ELUTING TOO SOON

1. Change pH to increase sample ionization - make pH similar to pKA.
2. Change buffer to contain ions less likely to compete for mobile ion on the packing.
3. Use stronger ion exchange packing.

Solvent polarity too strong
1. Reduce solvent strength
2. Change packing polarity
3. Ultimately consider Reverse Phase (if starting with Normal Phase) or vice versa.

EXCESSIVE RETENTION

1. Increase common ion in solvent.
2. Change pH to suppress ionization of the packing (weak exchangers only)
3. Use a weaker ion exchange packing.
4. Increase temperature.

1. Increase solvent strength
2. Reduce packing polarity

V_o

V_o

NO RESOLUTION

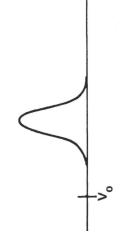

1. Reduce flow rate.
2. Use a longer column.
3. Reduce concentration of common or competing ion in the mobile phase.
4. Use different packing with comparable ionization properties.
5. Change from acidic to basic (or vice versa).
6. Consider non-ionic affinity separations.

1. Reduce flow rate
2. Use longer column or recycle
3. Try another solvent of comparable polarity but different hydrogen bonding capacity, functionality.
4. Use a new packing with about the same polarity as the previous one but with a different functional group.

MARGINAL RESOLUTION

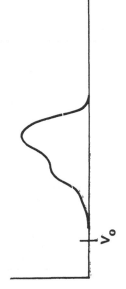

1. Reduce flow rate.
2. Increase length of column.
3. Reduce common ion concentration.

1. Reduce flow rate
2. Use longer column or recycle
3. Reduce solvent strength for sample.

FIG. 9. Steps to a satisfactory separation — affinity mode.

NO COMPONENTS RETAINED

Packing exclusion limit too low, with all
sample components excluded and eluting at
the solvent front.
Use a higher exclusion limit packing.

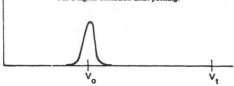

NO COMPONENTS EXCLUDED

Packing exclusion limit too high. Use a
lower exclusion gel.

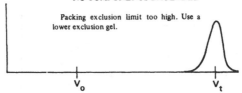

SOME ELUTION IN MORE THAN
ONE COLUMN VOLUME

Some components adsorbing on the packing. Use a stronger solvent system to insure
exclusion as the only operative mode.

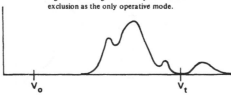

RESOLUTION MAY NOT BE
ADEQUATE

Use more columns with the exclusion
limit corresponding to the retention zone in
this chromatogram. Consider recycle.

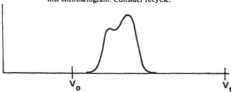

FIG. 10. Steps to a satisfactory separation — gel permeation mode.
Here V_o is the interstitial volume, V_t is the total column void volume (all
components will elute within V_t in true gel-permeation chromatography),
and $V_t - V_o$ is the packing pore volume.

C. Steps to a Satisfactory Separation

The chromatograms most likely to occur after the first try are shown in Figs. 9 and 10. Steps to take in each instance to achieve satisfactory resolution are listed with each chromatogram in the recommended order. As the desired degree of separation begins to develop, the changes will become more subtle. In refining the affinity separation two- and three-component solvent mixtures are often used. In these cases slight polarity shifts might be brought about by varying isopropanol content, for example, instead of methanol, since the latter solvent would cause more extreme fluctuations in solvent polarity than would be necessary in the latter stages.

III. CONTROLLING RESOLUTION

A. Dealing with Excessive Resolution — Solvent Programming

In some LC separations early eluting components may be only marginally resolved. Other components may take longer to elute than necessary, causing excessive length of analysis time and undesirable peak broadening. These situations have been termed the "general elution problem" [2].

To improve resolution of early eluting peaks, flow rate and solvent strength of the mobile phase should both be reduced. Reducing retention of the late-eluting peaks requires just the opposite. Sometimes a single combination of solvent and flow rate cannot satisfy these divergent requirements. It is then necessary to change conditions during the separation. The most pronounced effects on resolution occur when changing solvent flow rate or solvent composition (strength).

1. Flow Programming

Changing the solvent flow rate is flow programming. It is performed by controlling one solvent delivery system with a programmer to deliver a flow profile suitable for reducing retention times of certain components (Figs. 11 and 12).

Flow programming offers a number of benefits. The column is always in equilibrium with a single solvent, and no retrace of the program is ever required for column reequilibration. The differential refractometer can be used with flow programming, and the inherent reproducibility of this method makes it feasible for repetitive quantitative control use.

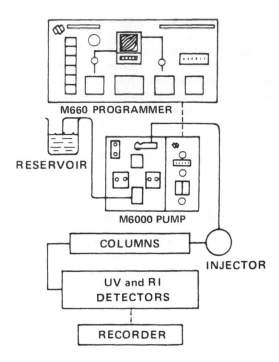

FIG. 11. Flow-programming schematic.

2. Gradient Elution

Changing solvent strength is solvent programming, or gradient elution. In solvent programming two solvent delivery systems are controlled simultaneously with a programmer to deliver a specific change in solvent composition during the program (Fig. 13).

Solvent programming can produce the greatest changes in the elution of components. It is especially useful in ion-exchange separations of biologicals with components having widely different ionic and/or pH strengths. By programming either buffer concentration or pH as appropriate, all components will elute within realistic operating times. Nonionic solvent programming is performed by increasing the solvent strength from the beginning to the end of the separation (Fig. 14).

The detection of eluting components in solvent programming is accomplished with an ultraviolet photometer, providing none of the solvents absorb at the detection wavelength. At the completion of the solvent program, especially in normal-phase work, the gradient should be run in reverse to return the column to its initial equilibrium of mobile and stationary phases.

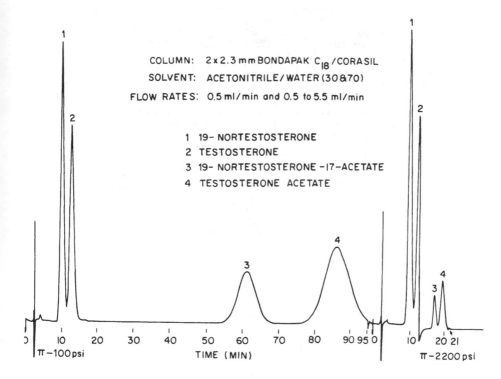

FIG. 12. Reduced analysis time via flow programming.

B. Dealing with Insufficient Resolution — Recycle

In some HSLC applications the maximum resolution attainable in a single pass through the system may not be sufficient to separate extremely similar compounds. To deal with this situation, many leading chromatographers have begun to employ recycle, a technique that greatly increases effective column length [3].

In recycle the sample is passed in series through the column and detector, then back into the pump inlet and through the column system and detector again. This is repeated as many times as necessary to achieve the desired amount of separation. A basic recycle schematic is included in Fig. 1.

The fundamental requirement for operating in recycle is a chromatographic system with very low peak spreading throughout the entire series of components and columns. Low peak spreading enables multiple passes of the sample through the system while preserving whatever partial resolution has already been achieved in the preceding passes.

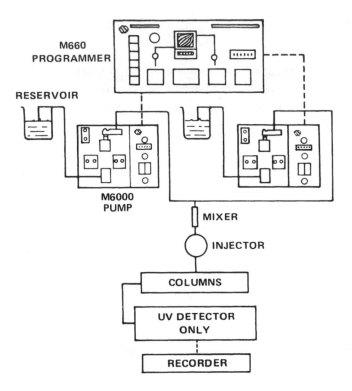

FIG. 13. Solvent-programming and scouting schematic.

Since resolution is increased by adding more columns or increasing the number of passes through the original columns, the advantages of using recycle are obvious:

1. Column costs are minimized

2. Fewer columns have to be handled

3. Pressure requirements remain modest and practical

4. In dealing with a new problem it may not be necessary to prepare or purchase additional columns to achieve good resolution

5. Solvent costs are minimized since the same solvent is recirculated; this is especially significant in preparative-scale work

It is often desirable to determine the purity of a compound that elutes as a single peak. In cases where extensive screening of solvent/packing combinations has already been performed without any evidence of a second component, recycle represents the ultimate test of the purity of the sample.

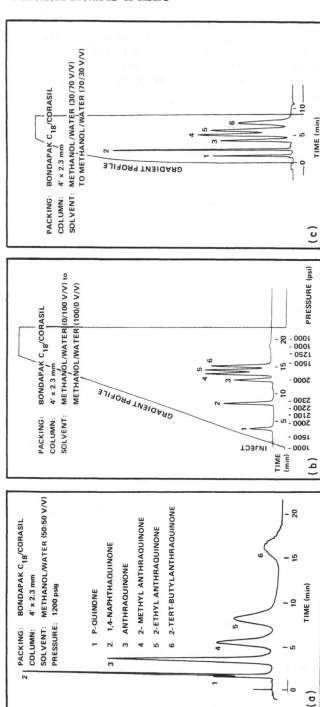

FIG. 14. Refining a reverse-phase separation with solvent programming. In the constant composition separation (a), components 1 and 2 are not fully resolved, component 6 is retained excessively, and the duration of the run is too long. In the first attempt to refine the resolution and reduce the total analysis time (b), the linear gradient gave excess resolution of the early peaks and did not reduce the total time appreciably. The unusual pressure readings are a consequence of the pronounced viscosity changes as the water-methanol ratio is varied. Satisfactory resolution of all components and a significant reduction in analysis time was achieved (c) by starting with a stronger solvent and using a convex gradient. If the problem involves fewer components, further reductions in analysis time are feasible by adjusting starting or ending solvent strength and gradient curvature.

Even if there is no apparent resolution after multiple recycles, early-
eluting and late--eluting fractions of the peak can be collected. These
fractions can then be compared by nuclear magnetic resonance, mass
spectrometry, or ultraviolet or infrared spectroscopy to verify their
identify or identify differences.

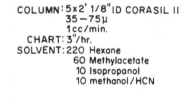

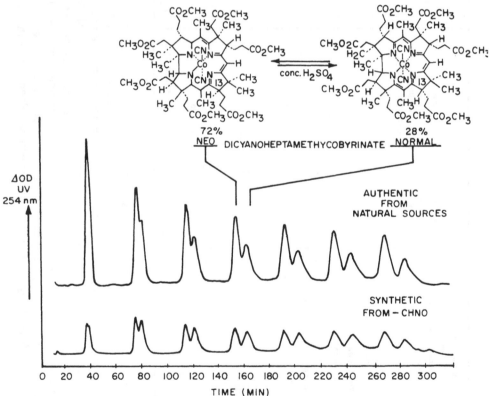

FIG. 15. Separation of the neo and normal forms of cobester by
recycle liquid chromatography.

FIG. 16. A complete liquid-chromatography system.

The resolving power of recycle is indicated in Fig. 15. The only difference between the neo and normal forms of dicyanoheptamethycobyrinate is the d and l configuration of the carbon in the 13-position. Yet base-line resolution is achieved in only seven passes through the system. This work was performed in the adsorption mode of liquid chromatography on 10 ft of 1/8-in. i.d. narrow-bore column. "Equivalent" resolution in a single pass would have required 70 ft of column!

This work was performed by R. B. Woodward and co-workers at Harvard University during their synthesis of vitamin B_{12} [4]. The agreement in chromatographic behavior between the isomer mixture derived from natural sources and the synthesized isomers verified the identity of the synthetic and naturally occurring mixtures.

Figure 16 shows a complete high-speed liquid chromatograph containing ultraviolet-photometer and differential-refractometer detectors. It is fully equipped for flow programming and solvent programming, and is also capable of recycle operation.

IV. PREPARATIVE LIQUID CHROMATOGRAPHY

A. Extending the Separation Spectrum

The preparation of pure compounds depends on the use of effective separating methods. Modern high-speed liquid chromatography is an exceptional complement to traditional separation techniques, increasing the resolving capability available to the investigator. The ease of resolving closely related compounds by modern liquid chromatography has, in fact, established higher standards of purity than were considered practical with previous methods.

All modes of high-speed liquid chromatography can be scaled up to produce ultrapure materials. Preparative liquid chromatography can be performed to isolate a few milligrams of materials or to prepare gram quantities, depending on the intended use of the collected material. The scope of preparative liquid chromatography includes the following:

1. Preliminary cleanup: removing large quantities of extraneous materials prior to gas chromatography or high-speed liquid chromatography

2. Analytical support: preparing pure standards

3. Synthesis support: determining purity of starting material; studying reaction mechanisms and kinetics; optimizing yield; isolating side-reaction products for identification

4. Preparing milligram to gram quantities of ultrapure materials for biological testing

5. Commercial preparation of rare specialty chemicals and biologicals

B. How To Do Preparative Liquid Chromatography

The preparation of pure samples may be approached as follows:

1. Defining the problem

2. Preparing the sample for liquid chromatography

3. Developing the analytical separation

4. Scaling up to preparative level operation

To scale up the analytical separation, additional sample-loading capacity is achieved in two ways: fully porous packings (Fig. 17) are substituted for the high-efficiency, low-capacity pellicular packings (Table V), and the solvent polarity is adjusted to expand the resolution. The latter approach moves the peaks farther apart to accommodate the peak widening that occurs when the active sites on the packing become overloaded as sample size is increased.

FULLY POROUS
HIGH CAPACITY

PELLICULAR
HIGH EFFICIENCY

PORASIL
STYRAGEL
PORAGEL
DURAPAK/PORASIL

CORASIL
DURAPAK/CORASIL
BONDAPAK/CORASIL

FIG. 17. Packing structures.

The overall preparative sequence is indicated in Fig. 18. Once the loading capacity is established, but before scaling up to the final operating level, the fractions of interest should be collected and run on the analytical scale, including recycle, as a check on their purity. Then Table VI should be used to select the column dimensions required to produce the desired amount of material. It should be noted that doubling the column length, with all other parameters unchanged, will at least double the sample-loading capacity.

TABLE V

Analytical and Preparative Packings

Analytical (pellicular)	Preparative (fully porous)	Surface polarity
Corasil	Porasil	High
Durapak Carbowax 400/Corasil	Durapak Carbowax 400/Porasil	Intermediate
Bondapak C_{18}/Corasil	Bondapak C_{18}/Porasil	Low

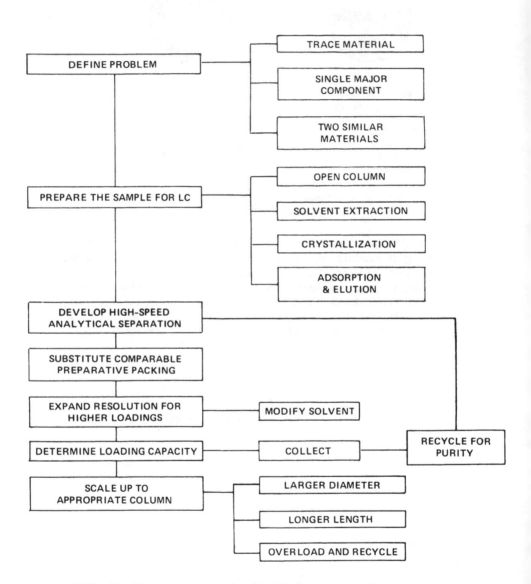

FIG. 18. Steps in preparative liquid chromatography.

TABLE VI

Preparative LC Loading[a]

	Affinity Mode				Gel Permeation Mode		
Column outside diameter (in.)	1/8	3/8	1	2.4	3/8	1	2.4
Relative internal cross-sectional area	1×	12×	100×	750×	1×	8×	60×
Typical column length (ft)	8	8	8	8	16	16	16
Typical sample load (g) – easy separation (only possible by TLC)	0.04	0.5	4	40	2	18	120
Typical sample load (g) – difficult separation ($\alpha < 1.3$)	0.004	0.05	0.4	4	1	9	60
Typical injection volume (ml)	0.005–0.01	0.5–4	4–10	10–100	5–10	45–90	300–600
Typical solvent flow rate (ml/min)	0.3–3	1–10	5–90	60–600	0.5–3	3–30	20–200

[a]Basis: fully porous packings

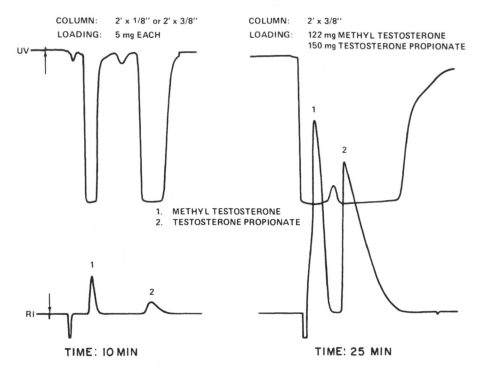

FIG. 19. Steroid loading studies.

C. Column-Loading Considerations

A systematic determination of column-loading capacity is performed by injecting increasing amounts of material in successive runs. Peak height will increase with higher sample loading until the active sites of the stationary phase become overloaded. Then the peaks will broaden and will ultimately overlap any adjacent peaks that are present.

In the loading study shown in Fig. 19, 5 mg each of methyl testosterone and testosterone propionate were injected onto 1/8-in. o.d. × 2 ft and 3/8-in. o.d. × 2 ft columns packed with Bondapak C_{18}/Porasil, a preparative, reverse-phase packing material. Comparable resolution occurred at both column diameters, as expected. The results obtained with the 3/8-in. o.d. column are shown in the chromatogram on the left. The separation was monitored with ultraviolet-photometer and differential-refractometer

detectors. Subsequent increases in loading led to the right-hand chromatogram, 150 mg of testosterone propionate and 122 mg of methyl testosterone.
The benefit of dual detectors is shown in this situation.

Although the ultraviolet absorption of small amounts of these steroids
was sufficiently strong to virtually blind the ultraviolet detector, it is
apparent from the refractometer trace that on a preparative basis resolution
of these two components is still considerable. Despite the apparent peak

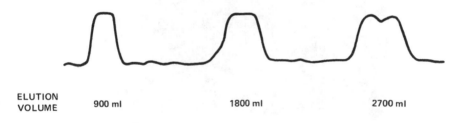

COLUMN: 3/8'' OD x 9' PORASIL T
SOLVENT: 1% ISOPROPANOL IN HEXANE (V/V)
SAMPLE: 100 mg IN 2 ml LOOP
INSTRUMENT: ALC-100 WITH UV PHOTOMETER AT 0.64 OD

ELUTION
VOLUME   900 ml   1800 ml   2700 ml

3600 ml   4500 ml

FIG. 20. Preparative separation of stereoisomers by recycle liquid
chromatography. From Koreeda, Weiss, and Nakanishi [5].

FIG. 21. Commercial preparative liquid chromatograph.

merging indicated by the ultraviolet photometer, each peak was better than 99% pure by other tests. Additional loading could still be tolerated before gross overlapping occurred. The importance of matching the detector characteristics to the applications is vividly demonstrated in this instance.

A dramatic example of preparative-scale recycle was performed by Nakanishi and co-workers at Columbia University during the determination of the absolute configuration of (+)-abscisic acid, an important plant hormone [5]. At one point in the study it was necessary to separate diastereomers into pure fractions. This was accomplished with 100 mg of the mixture in only five passes through a system containing a 9-ft column, as shown in Fig. 20. Equivalent single-pass separation would have required 45 ft of column.

A system designed specifically for preparative-scale high-pressure liquid chromatography is shown in Fig. 21. It has provision for injection of large sample volumes, high solvent flow, recycle, and automatic fraction collection — all essential for extensive preparative work.

V. SYSTEM CONSIDERATIONS FOR VERSATILE LIQUID CHROMATOGRAPHY

Any high-pressure liquid chromatograph must have the basic components shown in Fig. 1. There are specific requirements imposed on each component for satisfactory, full-range operation:

1. The solvent supply should be unlimited for long-term operation with large-diameter columns. It should be external and capable of being varied from a 1-liter solvent bottle to a large carboy, depending on the scale of operation.

2. The pumping system should be operative with an external solvent supply and thus not volume limited, but capable of prolonged unattended operation. It should have a low displacement volume for preparative injection through the pump, for recycle operation, and for rapid solvent changeover. It should also have a wide flow-rate range for convenient operation with small- or large-diameter columns; rapid flow-rate setting for convenient solvent changeover and injection of large samples at high flow rates; high-pressure capability for operation with long columns; adjustable pressure relief to match the column system in use; and constant flow for reliable, reproducible analyses.

3. Sample introduction should accommodate microliter or milliliter samples and should be compatible with highly concentrated, viscous solutions.

4. Columns should have high loading capacities and broad selection of column diameters to accommodate all operating scales.

5. Column-packing materials for all modes, including reverse phase, should be available. Comparable analytical (pellicular) and preparative (fully porous) packings should be on hand. In addition, the packing materials should be economical, durable, and stable.

6. Detectors should be flow insensitive, universal, sensitive, capable of handling high concentrations of major components, selective, and stable.

TABLE VII

Complete Liquid Chromatography

Modes	Capability	Scale
Gel permeation	Sequential analysis	Analytical
Ion exchange	Flow programming	Preparative
Liquid-liquid	Gradient elution	
Liquid-liquid bonded phase	Recycle	
Liquid-solid		

VI. CONCLUSION

Modern high-speed liquid chromatography provides a wide range of separation capability. It is an effective complement to analytical methods, wet or spectroscopic, as well as bioassay or synthesis work that relies on the availability of the pure fractions that liquid chromatography can provide.

Furthermore, the scope of available modes makes the use of two or more LC modes in sequence ideal for the analysis of complex mixtures [6]. Performing a size separation on a complicated mixture of compounds gives fractions that are much more manageable for subsequent analysis by other modes. This technique of sequential analysis is especially appropriate for evaluating plastics or natural products that contain a broad distribution of high- and low-molecular-weight compounds.

The complete capabilities matrix listed in Table VII indicates the many options that modern liquid chromatography offers for handling the most challenging separation problems.

REFERENCES

1. G. J. Fallick and J. L. Waters, Am. Lab., 4(8), 21 (1972).

2. L. R. Snyder, Principles of Adsorption Chromatography, Dekker, New York, 1968, Chapter 2.

3. K. J. Bombaugh, W. A. Dark, and R. N. King, Res. Dev., September 1968, p. 28.

4. T. Maugh, Science, 179, 266 (1973).

5. M. Koreeda, G. Weiss, and K. Nakanishi, J. Am. Chem. Soc., 95, 239 (1973).

6. J. N. Little, Am. Lab., 3(12), 59 (1971).

Chapter 4

MEASUREMENT OF DIFFUSION COEFFICIENTS BY GAS-CHROMATOGRAPHY BROADENING TECHNIQUES: A REVIEW

Virgil R. Maynard* and Eli Grushka

Department of Chemistry
State University of New York at Buffalo
Buffalo, New York

*Present address: Analytical Research and Services Laboratory, Central Research Laboratories, 3M Company, St. Paul, Minnesota

I. INTRODUCTION

Knowledge of diffusion coefficients is important in many areas of both basic and applied research. Diffusion plays an important role in chemical reactions and must be considered in the design of distillation columns. Diffusion also plays a major role in peak broadening in chromatography, and accurate values of diffusion coefficients are often necessary in the testing of chromatographic theory.

Mason and Marrero [1] and later Marrero and Mason [2] have written excellent reviews on gaseous diffusion in general and all the known methods of obtaining such data, including the gas-chromatography (GC) method. This method, first introduced by Giddings [3], has been used by many other workers both as originally conceived by Giddings and in several modified forms, as will be seen in Sections III and IV. It is our purpose to critically review the literature concerning the measurement of gas-gas and gas-liquid vapor binary diffusion coefficients by methods based on the chromatographic broadening technique.

II. THEORY

A. Mass-Balance Equation

The problem of diffusion in flowing fluids was first studied extensively by Taylor [4-6] and by Aris [7] in the 1950s. From their work it can be seen that the diffusion of a trace amount of a solute in an open tube containing a flowing solvent can be explained by using a mass-balance approach to the problem. The mass-balance equation that applies here is

$$\frac{\delta c}{\delta t} - D_{AB}\left[\frac{\delta^2 c}{\delta x^2} + \frac{1}{r}\frac{\delta}{\delta r}\left(r\frac{\delta c}{\delta r}\right)\right] + 2\bar{U}\left[1 - \left(\frac{r}{r_0}\right)\right]^2\frac{\delta c}{\delta x} = 0 \tag{1}$$

where c is the concentration, t is the time, D_{AB} is the binary diffusion coefficient of the solute-solvent pair, x is the longitudinal coordinate of the tube, r is the radial coordinate of the tube, $\bar{U}$ is the average solvent (i.e., carrier gas) velocity, and r_0 is the radius of the tube. To solve this equation, we must establish boundary conditions and make certain simplifying assumptions.

1. None of the solute can pass through the tubing wall, that is, $(\delta c/\delta r)_{r=r_0} = 0$

2. At the center of the tube the radial concentration gradient is zero, that is, $(\delta c/\delta r)_{r=0} = 0$

In addition, we must assume that the solute is introduced as a δ function, that the solute does not interact with the wall (i.e., no adsorption effects), and the ratio of solute-wall collisions to solute-solvent collisions is small. Also, we must assume that turbulence is absent and the flow is laminar. Given these boundary conditions and simplifying assumptions, Eq. (1) can be solved to give

$$C = A\left(\frac{D_{eff}\,t}{L}\right)^{-1/2} \exp\left[\frac{-(1 - \bar{U}t/L)^2}{4(D_{eff}/\bar{U}L)\,(\bar{U}t/L)}\right] \tag{2}$$

where A is a constant that depends on the amount of solute injected, L is the length of the tubing, and D_{eff} is an effective diffusion coefficient given by [2]

$$D_{eff} = D_{AB} + \frac{r_0^2\,\bar{U}^2}{48D_{AB}} \tag{3}$$

The first term is the diffusion coefficient and describes broadening of the band in the axial direction due to the concentration gradient in that direction. The second term, the Taylor diffusion coefficient, takes into account band broadening due to the parabolic flow profile and to radial diffusion.

Equation (2) describes a skewed Gaussian. In the limit of a long column and a slow carrier-gas velocity $(D_{eff}/\bar{U}L \leq 0.01)$ Eq. (2) becomes a Gaussian whose variance in length units is

$$\sigma^2 = \frac{2D_{AB}L}{U} + \frac{r_0^2\,\bar{U}L}{24D_{AB}} \tag{4}$$

In chromatography the plate height H is defined by

$$H = \frac{\sigma^2}{L} \tag{5}$$

Substitution of Eq. (5) into Eq. (4) yields

$$H = \frac{2D_{AB}}{\bar{U}} + \frac{r_0^2\,\bar{U}}{24D_{AB}} \tag{6}$$

B. Golay Equation

Alternatively we can employ the Golay equation, which describes band broadening in coated open-tube columns [8, 9]:

$$H = \frac{2D_{AB}}{\bar{U}} + \frac{2R(1-R)}{3} \frac{d_f^2 \bar{U}}{D_L} + \frac{(11 - 16R + 6R^2)r_0^2 \bar{U}}{24D_{AB}} \tag{7}$$

Here R is the ratio of the solute velocity to the carrier-gas velocity, d_f is the thickness of the stationary-phase film coated on the tube, and D_L is the diffusion coefficient of the solute in the stationary phase. Since we are concerned here with the case where there is no coating on the tube ($d_f = 0$), and where there is no retention of the solute ($R = 1$), the Golay equation reduces to Eq. (6). The same result is found if one employs either the van Deemter equation [10] or the Giddings coupled equation [8]. Equation (6) can be rearranged to yield

$$D_{AB} = \frac{\bar{U}}{4}\left[H \pm \left(H^2 - \frac{r_0^2}{3}\right)^{1/2}\right] \tag{8}$$

Equation (8) gives two values for the diffusion coefficient, only one of which is meaningful. When the velocity is slow, the second term on the right-hand side of Eq. (6) is small and D_{AB} is determined from the positive root. At high velocities the first term is small and the negative root is used to calculate the correct diffusion coefficient. The cross over point from one root to the other is at the velocity, $\bar{U}_{opt}$, that minimizes H. Typical behavior of Eq. (8) is illustrated in Fig. 1. The velocity $\bar{U}_{opt}$ can be found by differentiating Eq. (6) with respect to $\bar{U}$, setting the answer equal to zero, and solving for $\bar{U}_{opt}$. If this is done, then

$$\bar{U}_{opt} = \frac{\left(48D_{AB}\right)^{1/2}}{r_0} \tag{9}$$

Some workers [11] have determined D_{AB} from Eq. (9) by experimentally determining $\bar{U}_{opt}$. To solve Eq. (8), we must know r_0, $\bar{U}$, and H. The value of r_0 can be measured directly, and $\bar{U}$ can be found by dividing the length of the tube, L, by the retention time t_R of the solute, both of which can be measured directly. The value of H can then be obtained experimentally as

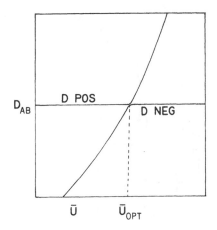

FIG. 1. Typical behavior of the parameters in Eq. (8).

$$H = \frac{LW_{1/2}^{2}}{5.545t_{R}^{2}} \tag{10}$$

where $W_{1/2}$ is the width of the peak at one-half the peak height.

III. EXPERIMENTAL

A. Apparatus

In this section we shall illustrate the various experimental methods for measuring diffusion coefficients by gas chromatography. The most basic setup, shown in Fig. 2, is a commercial GC apparatus where the packed column is replaced with a coiled, long, empty tube of circular cross section [3].

Most work has been done with open tubes and conventional apparatus, but Arnikar, Rao, and Karmarkar [12, 13] have used packed columns and an electrodeless discharge detector. More will be said later in Section IV on packed-column work.

Some authors have used specially built equipment that bears little resemblance to a gas chromatograph. For example, Bournia, Coul, and Houghton [14] have built an apparatus containing a large cylindrical sample chamber. The carrier flow rate was regulated with a bubble-type flow regulator. After purging the chamber with the sample gas, injection into a

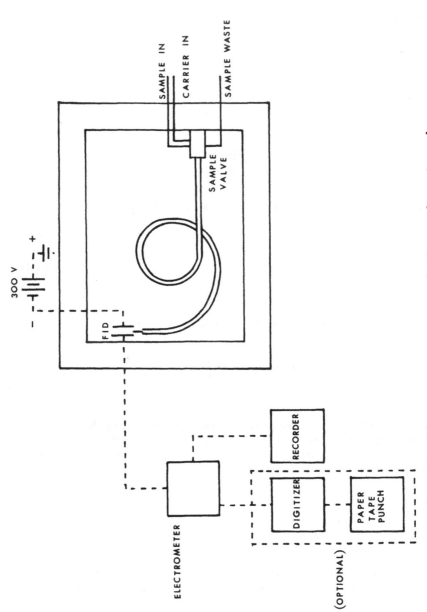

FIG. 2. Basic apparatus for measuring diffusion coefficients by gas chromatography.

Vycor-glass tube was accomplished by simultaneously releasing two slide valves. A hydrogen discharge lamp emitted ultraviolet radiation from which a Beckman D.U. spectrophotometer provided a narrow band of 250-nm wavelength. This light passed through the tube and impinged on a phototube. The current produced flowed through a large resistor, and the resultant voltage drop was amplified and displayed on a strip-chart recorder.

Evans and Kenney [15] used a somewhat similar apparatus, except they used a much longer tube made from steel, a thermal conductivity detector, and a six-port injection valve.

B. Sources of Errors

One of the problems associated with diffusion-coefficient measurement is the effect of finite injection volume, detector volume, and dead volume due to connecting tubing. To correct for this, Giddings and Seager [3] have used two columns. The first, the analytical column, was 30 m long whereas the other, a short correction column, was 1 m long. The correction for extracolumn band spreading was based on the assumption that all of the contributions to peak spreading are independent of one another. One can then subtract the variance of the peak obtained with the 1-m column from the variance of the peak obtained with the 30-m column and get the variance of a peak that would be produced in a 29-m column with no end effects. This procedure is absolutely necessary when any significant amount of dead volume is present in a system.

Ideally, the sample would enter the column as an infinitely narrow plug. This can be difficult to approach experimentally. Ingenious injection devices have been built by several authors [14, 16-19]. Although they all differ in detail, their basic design is such that the sample gas or vapor flows continuously through one part of the device and the carrier gas flows through another. At the moment of injection a portion of the sample is trapped and then placed in the carrier stream.

Whatever variations on the basic apparatus are used, careful attention must be paid to (1) ensuring that temperature gradients are kept to a minimum in the vicinity of the column and (2) providing a steady flow of the carrier gas. The problem of temperature gradients can be eliminated through careful arrangement of the oven geometry [20] or using a liquid constant-temperature bath [21]. Eliminating variations in the carrier-gas flow can be a difficult problem with gas tubes since there is virtually no pressure drop across the column. One way of eliminating this problem is to put a tight constriction in the carrier-gas line just before the point where a sample is injected. Carrier-gas flow irregularities are most apt to be a problem when injection valves are used since the steady flow of carrier gas is interrupted at the moment of injection. These injection valves are desirable, however, because injection volume is far more reproducible and the injection profile more closely approximates a δ function.

Other sources of errors are due to column coiling (the so-called race-track effect), the secondary flow phenomenon, and stagnant pockets. Careful instrumentation design (i.e., large coil diameter, slow flow rates, and minimum dead volume) can greatly diminish these difficulties [22]. Adsorption of the solutes on the column wall can also contribute to the peak width. It is advisable, therefore, to use tubes containing low-energy adsorption sites and small ratios of wall surface to volume.

IV. DISCUSSION

A. Comparison with Other Techniques

1. Closed Tube

The closed-tube technique was developed by Loschmidt [23, 24] in 1870. The apparatus consists of a long tube closed at both ends with a fast opening valve in the middle. Samples of the pure gases are initially separated and then allowed to mix by diffusion, and the composition of each section is determined after a period of time. This apparatus yields excellent values for gas-gas pairs. The principal disadvantage is the relatively long analysis time. The precision and accuracy of the method are quite good, better than those obtained with all but the most precise chromatographic methods.

2. Evaporation Tube

Stefan [25] developed the evaporation-tube method for measuring the diffusion coefficients of liquid vapor-gas mixtures. In this method a liquid or volatile solid is placed in the bottom of a short tube and the loss of the material through evaporation into the gas is measured. This method, however, has poor precision (greater than 5%), and measurements take half a day to perform. Here the GC method is seen to be vastly superior.

3. Two-Bulb Apparatus

The two-bulb apparatus developed by Ney and Arnistead [26] is essentially an improvement of the closed-tube method. Two bulbs containing the two diffusing gases are joined by a narrow tube. The device is more compact than the closed-tube apparatus and hence easier to thermostat. The precision and accuracy of the two-bulb and closed-tube devices are about the same.

4. Point Source

The point-source method of Westenberg and Walker [27] is very similar
in principle to the GC method, except that the tracer is continuously steam-
ing into the flowing carrier. The concentration of the trace gas is then
measured at different distances along the column. Using this method,
diffusion coefficients have been measured at temperatures of up to 1900° K.
The accuracy is average (about 5%). Marrero and Mason [2] have discussed
these methods in more detail. They have also reviewed many other methods
which can be (but seldom are) used for diffusion-coefficient measurement
and which we shall not discuss here.

The GC method has, over the years, been at worst a method with
average precision and accuracy. However, the method has few intrinsic
errors and, as we shall show in the next subsection, can be refined to have
a precision of 1 or 2%. This makes the GC method as accurate as any other,
and, being a dynamic method, it is faster than most other methods.

B. Chronological Review of Literature

Giddings and Seager [3] are generally given credit for publishing the first
paper on the use of gas chromatography to determine diffusion coefficients.
Their apparatus was described in Section III. They measured the diffusion
coefficient of hydrogen in a nitrogen carrier at 21 different velocities under
16 cm/sec at 293°K. Above 16 cm/sec the plot was no longer linear and
horizontal. The precision of their measurement was estimated to be about
2% and deviated from the literature value quoted by about 5% (see Appendix
Table I). This paper is important not so much for the data reported but
because it shows the potential of the technique as a fast and accurate method
of determining gaseous-diffusion coefficients. Giddings and Seager also
predicted that it may be used to calculate diffusion coefficients in liquids.

A paper published at practically the same time as Giddings's [3] was
one by Bohemen and Purnell [16], who measured diffusion coefficients at
very low velocities using unpacked columns and no correction tube (see
Appendix Table I). They ignored the Taylor diffusion coefficient in their
calculations because it contributed only about 3% to the band broadening
observed at the velocity at which they were working. However, when one is
doing precise work, there is little justification for not including the Taylor
coefficient. They have also obtained van Deemter "B" terms using packed
columns and unretained solutes. One column used contained Silo-O-Cel
firebrick coated with polyethylene glycol 400. The other column contained
glass beads. Bohemen and Purnell showed that the van Deemter equation
reduces to $H = A + B'/U'$, from which D_{AB} can be found from a plot of H
versus $1/U'$ ($B' = 2\gamma D_{AB}$, γ being the destructive factor, and U' = outlet
velocity at 1-atm pressure).

Although they apparently saw no connection with chromatography, Bournia, Coull, and Houghton [14] published a paper utilizing essentially the same technique as the previous two papers discussed. Their intention was to verify experimentally the work of Taylor for gas dispersion. In operation, however, their system cannot give results as precise as desirable for several reasons:

1. The sample volume is far too large, being about 8% of the total column volume. Thus its contribution to H would be excessive.

2. In addition to being large, the sample was also a pure gas. Thus it was impossible to measure diffusion coefficients for systems in which the solute and solvent varied greatly in density.

3. Up to the moment of injection $\bar{U} = 0$. Therefore, there must have been some fluctuations in velocity during an experimental run.

However, the work of Bournia et al. is still valid in that they have demonstrated that the Taylor theory of dispersion of a fluid flowing in a tube applies to gases as well as liquids in the velocity range of 2 to 15 cm/sec.

Fejes and Czaran [17] have written an interesting paper in that they derived diffusion equations useful in frontal analysis. This paper was published independently and almost simultaneously with the previous three papers on diffusion-coefficient measurements. The authors were doing frontal GC investigations on equilibria and kinetic relationships concerning adsorption. For this work they needed a better knowledge of diffusion in open and filled tubes. Their experimental setup was that of a typical gas chromatograph, except for the usage of a four-way valve that permitted them to switch from a stream of one gas to another, thereby making a step-function injection. Fejes and Czaran derived diffusion equations that are based on the work of Taylor and Golay, and account for convection, radial diffusion, and longitudinal diffusion. Their equations, however, are derived for the case of frontal analysis and are not directly usable for elution analysis, the more generally used technique. They presented data obtained at a temperature of $298^\circ K$ and 1-atm pressure. The gas pairs studied were (1) nitrogen, methane, ethane, propane, and n-butane in hydrogen; (2) ethane, propane, and n-butane in nitrogen; and (3) methane in carbon dioxide. The data were presented with only two significant figures but agreed well with the literature values quoted (see Appendix Table I).

Giddings and Seager's [28] second paper on diffusion-coefficient measurement was more detailed and included more data than the earlier communication. They noted especially the speed of the method, having made 200 separate determinations in 36 h while maintaining a precision of about 1%. With this accomplishment it has become possible to compile extensive

diffusion-coefficient measurements. This would be quite difficult to do
with the older, more time-consuming, methods. The authors point out,
again, that the method is equally suitable for measuring diffusion coefficients in liquids and gases. Diffusion coefficients for argon, hydrogen,
ammonia, nitrogen, and oxygen in helium; hydrogen and carbon dioxide
in nitrogen; and carbon dioxide in hydrogen were presented in tabular
form (see Appendix Table I). When the carrier gas and trace component
were reversed, corresponding to the gas pairs nitrogen-helium and
helium-nitrogen, the two values obtained bracketed the values obtained
with a Loschmidt apparatus. Concentration effects were determined by
injecting the trace component into a mixture of the carrier gas containing successively higher and higher concentrations of that trace component.
For the helium-carbon dioxide system the diffusion coefficient increased
slightly with the mole fraction of helium in the carrier, whereas the
opposite trend was noticed in the helium-nitrogen system.

Knox and McLaren [11] determined the diffusion coefficient of the
nitrogen-ethylene system to help prove a HETP equation they were proposing. They investigated the two generally accepted but differing
equations for HETP: the van Deemter equation,

$$H = A + \frac{B}{U} + C\bar{U} \tag{11}$$

and the coupled equation of Giddings,

$$H = \frac{B}{\bar{U}} + C_1\bar{U} + \left(\frac{1}{A'} + \frac{1}{C_g\bar{U}}\right)^{-1} \tag{12}$$

These authors combined the two equations, thus obtaining the following
HETP equation:

$$H = A + \frac{B}{\bar{U}} + \left(\frac{1}{A'} + \frac{1}{C_g\bar{U}}\right)^{-1} + C_1\bar{U} \tag{13}$$

They then sought to test their equation. To simplify matters, they used a
nonsorbed solute, thus setting C_1 equal to zero. The diffusion coefficient
was then found by using a capillary column. The value of $\bar{U}_{opt}$ was found
experimentally, and D_{AB} was then calculated from Eq. (9) (see Appendix
Table I).

Seager, Geertson, and Giddings [29] investigated the temperature
dependence of gas-gas and gas-liquid vapor diffusion coefficients. This
was the first paper in which the diffusion coefficients of gas-liquid vapor

systems were evaluated. Diffusion data were given for four gas-gas pairs (nitrogen, oxygen, argon, and carbon dioxide in helium) at eight temperatures ranging from 298 to 498°K (see Appendix Table I). As in their previous work [3, 28], a correction tube was used to compensate for end effects. The same apparatus used previously [3, 28] was employed. The liquid samples were benzene, methanol, ethanol, 1-propanol, 2-propanol, 1-butanol, 1-pentanol, and 1-hexanol, all with helium as the carrier gas. The 0.1-μl liquid samples were injected with a 1-μl syringe. The temperatures utilized in the gas-liquid vapor work ranged from 423 to 523°K. The diffusion coefficient was assumed to be proportional to T^m. The exponent m was evaluated for each system and varied rather widely.

The average value of m both for the four gas-gas pairs and for the eight gas-liquid vapor pairs was 1.70. The values for the diffusion coefficients were corrected to 1-atm pressure and agreed fairly well with literature values.

The value of GC determination of diffusion coefficients is illustrated well in a paper by Barr and Sawyer [19]. In their study on the liquid-phase mass-transfer term in gas chromatography the authors needed to know the diffusion coefficients of 3-pentanone in three carrier gases: helium, argon, and nitrogen. The diffusion coefficients were calculated from data obtained with a conventional gas chromatograph equipped with a 16-ft × 1/8-in. empty tube and a hydrogen flame ionization detector. The diffusion coefficients were calculated from Eq. (8) (see Appendix Table I).

Knox and McLaren [18] have developed a stopped-flow GC technique for determining gaseous-diffusion coefficients, D_{AB}, and also the obstructive factor γ. The method is as follows: when a vanishingly small plug of a gas A is allowed to diffuse into gas B, the concentration profile of A in B is Gaussian and the rate of spreading is determined by

$$\frac{d\sigma_x^2}{dt} = 2D_{AB} \tag{14}$$

If the system is moving at a constant linear velocity $\bar{U}$, the standard deviation in time units τ is given by $\tau = \sigma_x / \bar{U}$ and

$$\frac{d\tau^2}{dt} = \frac{2D_{AB}}{\bar{U}^2} \tag{15}$$

Knox and McLaren then determined the diffusion coefficient by injecting gas A as a sharp band into a moving stream of gas B in an empty column. The band was eluted about halfway down the tube where the carrier stream was stopped and A was allowed to spread into B, purely by diffusion for a time t. The band was then eluted, and its concentration profile was determined with standard GC apparatus. Equation (15) describes the variance added to the peak during the delay time t. The experiment was then repeated for the same velocity and different delay times t. The total variance was then plotted against the delay time. This yielded a straight line whose slope was $2D_{AB}/\bar{U}^2$ (see Appendix Table I). The experiment was repeated with packed columns. Equations (14) and (15) still apply, except that D_{AB} must be replaced with γD_{AB}. Using this method, which the authors described as the arrested-elution technique, D_{AB} and γ were determined for the system ethylene in nitrogen.

The arrested-elution method has one major point in its favor. Band broadening due to flow irregularities is held constant throughout the experiment and is effectively canceled out. Also, there is no reason why the column cannot be fairly short since it is required to be only long enough to contain the diffusing band. The authors state that there was little difficulty in arresting the flow so long as the pressure drop across the column was small (≤ 2 cm Hg). Their precision was about 2%. The main disadvantage of the method is the requirement of repeating the experiment at a variety of delay times of from 1 to 20 min.

Evans and Kenney [15] have confirmed Aris's solution for the mass-balance equation [our Eq. (1)] for the velocity region 1 to 16 cm/cm. For the nitrogen-ethylene system their experimental results agreed well with theory. For other systems the agreement with other literature values decreases with increasing difference in the molecular weights of the two species (see Appendix Table I). Their work was also undertaken to test equipment to be used to study mass transfer between gases and liquids. The apparatus consisted of a straight, drawn, mild-steel tube 1729 cm long, 3/8 in. o.d. by 1/4 in. i.d. formed by connecting 8-ft sections of tubing with bored-out brass couplings. The work was done at room temperature, and the authors depended on the tube's massive structure to eliminate temperature gradients. Injection of 1-ml samples was accomplished with a six-port injection valve. A twin-cell thermal conductivity detector was used to monitor the peak profile. An interesting sidelight of their experiments was the placement of plugs with 1/16-in. orifices at various points in the column. When this was done, there was a slight increase in the peak dispersion. Unfortunately Evans and Kenney did not pursue this study for the important case of a small orifice at the beginning of the column, the situation prevailing with some commercial injection valves.

Fuller and Giddings [30] have compared nine mathematical methods for predicting gaseous-diffusion coefficients. Although no experimental data were given, this paper is included in our review because it shows the difficulty of estimating diffusion coefficients from theory. The values of the diffusion coefficients estimated by the nine different equations were compared with experimental values for 38 systems, and percentage errors for each estimation were tabulated. It was found that the method of Fuller, Schettler, and Giddings (FSG) [31] gave the best estimate. Average absolute percentage errors varied from 4.2 to 20%, depending on the method employed. However, one characteristic of all the systems employed was a fairly unpredictable variation in the percentage error for individual cases. For example, the FSG method, while giving estimates that are low by 4.9 to 7.9% for most hydrocarbons in helium, estimates the diffusion coefficient of hexane in helium to be 17.4% lower than experimentally determined.

Huber and van Vught [32] have investigated the optimum working conditions for determining the diffusion coefficients by gas chromatography. After a discussion of the theory of solute dispersion in a fluid flowing in an open tube, conclusions are drawn concerning optimum working conditions. These conclusions were experimentally verified under varying opterating conditions. They determined that the diffusion coefficient, in either gas or liquid systems, could be determined with a precision of about 2% when the carrier velocity was adjusted to give solute retention times of about 10 min. The chief contribution to error in their system was the syringe injection system and the manual measurement of chromatograms. A more precise method of solute injection and data reduction will undoubtedly improve the precision (see Appendix Table I).

Chang [33] has determined diffusion coefficients for nitrogen-helium systems using a gas chromatograph of his own design. He has obtained diffusion data at pressures ranging from atmospheric to 900 psig and at temperatures of 244, 255, 277, 298, and 311°K. He also obtained diffusion data for trace amounts of ethane, propane, and n-butane in methane at 1-atm pressure and 255, 277, 298, and 311°K. The diffusivities compared well with literature data where available. The pressure variation of diffusion coefficients was compared with kinetic theory predictions, and good agreement was found.

Arnikar, Rao, and Karmarkar [12] have measured the diffusion coefficient of oxygen in nitrogen in a packed column (see Appendix Table I). The column was packed with silica gel. The main point of this paper, however, was to illustrate the usefulness of an electrodeless discharge tube as a GC detector. In a later paper [13] the same authors measured the diffusion coefficients of several gases and vapors, again in packed columns. In addition to the silica-gel column previously used, they also employed a column packed with 3% polyethylene glycol 1000 on 80-100-mesh firebrick. They used the van Deemter equation to calculate the diffusion coefficient. The columns were operated at very low carrier-gas velocities in which the van Deemter equation can be approximated by

$$H = A + \frac{2\gamma D_{AB}}{\bar{U}} \tag{16}$$

Arnikar and co-workers assumed that $\gamma = 1$ for their columns. Under this assumption a plot of H versus $1/\bar{U}$ yields a straight line with a slope of $2D_{AB}$. The authors again used their electrodeless discharge detector to obtain the peak profile. Their results were from 2 to about 14% higher than their quoted literature values. Their appears to be no real advantage to using packed columns. Indeed the assumption of $\gamma = 1$ is probably not valid and could account for the large errors (see Appendix Table I).

Giddings and Mallik [34] have reviewed several of the more "unorthodox" applications of gas chromatography. Among these are measurements of diffusion coefficients. The authors described the GC method briefly and reported two new diffusion coefficients for nitrogen-ethylene and nitrogen-butane at 302.6 and 302.4°K, respectively (see Appendix Table I).

Hargrove and Sawyer [35] have published diffusion-coefficient values of liquid vapor-carrier gas pairs at room temperature. The previous GC measurements of liquid vapor-gas diffusion coefficients were carried out at elevated temperatures. Comparison with work done by classical procedures necessitated extrapolation of the GC determination, and when this was done, the data differed by 8 to 10%. Experimental difficulties were encountered when measuring diffusion coefficients at 298°K. The solute vapor tended to adsorb to the tubing walls at this temperature. This difficulty was overcome by adding a constant amount of the solute to the carrier gas, thereby saturating the adsorption sites. Hargrove and Sawyer used Eq. (8) to calculate the diffusion coefficient. A modified commercial gas chromatograph with syringe injection of vapor samples was used. Values of D_{AB} at 1-atm pressure and 25°C for the systems helium-ethanol, helium-benzene, and nitrogen-n-butane were given and compared with literature values. The largest deviation was about 4% (see Appendix Table I).

In addition to determining D_{AB} at 298°K, Hargrove and Sawyer determined the diffusion coefficients for a variety of solutes at 372.6, 423.0, and 473.0°K. From the measured values of the diffusion coefficients it was possible to determine the temperature dependence of the diffusion coefficients using the equation

$$D_T = D_{ref}\left(\frac{T}{T_{ref}}\right)^{m} \tag{17}$$

where $D_T = D_{AB}$ at temperature T and $D_{ref} = D_{AB}$ at a reference temperature, T_{ref}. A plot of $\log D_T$ versus $\log (T/T_{ref})$ gave a straight line with a slope of m. The m values for the system were given and varied from 1.43 to 1.93.

Huang, Sheng, and Yang [36] have studied the GC method of determining diffusion coefficients with the intent of determining the idiosyncracies of the method. The diffusion coefficient was measured using our Eq. (8), with the plate height determined from an equation equivalent to our Eq. (10). Chromatograms were obtained with a commercial gas chromatograph using syringe injection and a thermal conductivity detector. Columns used included a U-shaped steel tube, a long coiled column, and a much shorter coiled column. The authors have calculated an effective dead volume for their apparatus that was found to be independent of the volume of the column, at least at very low carrier velocities. Very little data were presented, but from what there were there appears to be a velocity dependence of this dead volume. The authors found that the measured value of the diffusion coefficient varied with the velocity. For example, with the U-shaped column, a plot of D_{AB} versus $\bar{U}$ gave a straight line with a positive slope. The diffusion coefficient was independent of the velocity at very low carrier-gas velocities when the long coiled column was used but increased with velocity in a nonlinear fashion at higher velocities. The short coiled column exhibited similar behavior to the long coiled column, except that it had a smaller velocity-independent region and a more radical change in slope with increasing velocity. Huang et al. attributed the U-column behavior to flow irregularities at the injector-column junction. Eddy effects were claimed to be small for the coiled column due to the uniformity of the column, but the finite vaporization time of the liquid sample allegedly introduced a broadening effect. However, we wonder why finite vaporization was not a problem with the U-shaped column since liquid samples were injected here also. In any case, the diffusion coefficients increased with velocity for polar solutes (ethanol) and for nonpolar solutes (benzene). Huang et al. concluded that the most favorable conditions for measuring diffusion coefficients were a carrier velocity of about 1 to 1.5 cm/sec, trace sample size, and a column length of about 10 m. Binary diffusion coefficients of five hydrocarbons and four alcohols were determined in hydrogen with an average deviation of 2.7% (see Appendix I).

Zhukhovitskii, Kim, and Burova [37] have proposed two chromatographic methods for measuring gaseous diffusion coefficients. The first method involves saturating a capillary with helium and then connecting the capillary to another tube in which nitrogen was flowing. After a certain time t, the remaining helium was analyzed and the diffusion coefficient was determined from the following equation [38]:

$$\frac{C_{av}}{C_0} = \frac{8}{\pi^2} \exp\left(\frac{\pi D_{AB} t}{4L^2}\right) \tag{18}$$

where C_{av} and C_0 are concentrations of helium at times t and t_0. The main disadvantage of this method was the extremely long analysis time of 20 to 40 h. In the second method a sample of a weakly sorbed gas was allowed to diffuse into a column packed with carbon. After a certain time t, the weakly sorbed gas was displaced by a carrier and its concentration profile was plotted on a recorder. The diffusion coefficient was determined from an equation derived by the authors (see Appendix Table I):

$$D_{AB} = \frac{\mu^2 L}{11.08 svt^2} \tag{19}$$

where s is the cross section of the tube, μ is the peak width, and v is the retention volume.

Arnikar and Ghule [39] have measured the diffusion coefficients of several organic liquids in nitrogen using their electrodeless discharge detector and open tubes of circular cross section. They based their calculations on Eq. (6), recognizing three carrier-gas velocity regions. They did not use the high-velocity region, where the resistance to mass transfer term predominates, because the peaks were much too narrow. The intermediate-velocity region, where both terms are of similar magnitude, was also not utilized, probably because of the same experimental difficulties as in the high-velocity region. Arnikar and Ghule therefore worked at relatively low carrier-gas velocities, where

$$\lim_{u \to 0} H = \frac{2D_{AB}}{\bar{U}}$$

Arnikar and Ghule determined the diffusion coefficients of methane, ethane, isopropanol, methyl acetate, ethyl acetate, benzene, n-pentane, and acetone in nitrogen. The values obtained were uniformly higher than the literature values quoted, except for one that was compared to a GC-determined value in a packed column (see Appendix Table I). Actually, this is to be expected since their equation introduces a small error due to neglecting the mass-transfer term in the Golay equation. Also, their detector has considerable dead volume, which could contribute significantly to band broadening, hence increasing the apparent value of D.

Fuller, Ensley, and Giddings [22] have studied the diffusion characteristics of halogenated hydrocarbons in helium. Their objectives were (1) to develop improved experimental apparatus and techniques, (2) to determine the dependence of collision cross sections on molecular structure, and (3) to acquire reliable data for more diffusion pairs. The diffusion coefficients for 31 halogenated hydrocarbons diffusing into helium,

of which 29 were for new systems (see Appendix Table I). The relative
standard deviation of their work for most systems was about 2%. Fuller et
al. improved their apparatus by using a more spacious oven, permitting the
use of a column with a larger coil diameter and hence less distortion of the
sample profile. Their previous requirement of a short corrector column
was eliminated through the use of a gas-sampling valve permitting direct
on-column injection and a flame-ionization detector with very little dead
volume. Their gas-sampling valve injected 0.5 cm^3 of sample. Pure gases
were diluted with helium prior to sampling, whereas liquids were sampled
by bubbling helium through a steel chamber containing some of the liquid
and then injecting the solute-saturated helium. Fuller et al. have shown
theoretically that the broadening effect of their finite injection volume,
detector volume, stagnant pockets, and tube coiling was less than 0.5%.
They have expanded their list of special atomic diffusion volumes to include
fluorine, bromine, and iodine.

Of more fundamental importance is that, prior to the publication by
Fuller and co-workers, most work on collision cross sections was done with
simple molecules that approximated spheres or spherocylinders. This was
necessitated by the many possible rotational states, with unknown probability,
of complex molecules and the ensuing difficulty of measuring an effective
cross section. Also what was previously known about collision cross
sections was determined mainly from viscosity data and molecular-
beam scattering measurements. Binary diffusion measurement should
provide a better tool since it is a direct measure of interaction between
dissimilar molecules. Fuller et al. argue that, when helium is used as one
member of the diffusing pair, one has a sensitive probe into the cross
section of a much larger molecule. Helium is small, inert, and has a
spherical force field. It is thus an ideal probe for investigating changes in
cross section due to addition or rearrangement of specific groups within a
molecule. Their experimental data showed that the location of a halogen
atom in the 1-position resulted in smaller diffusion coefficients than with
the atom in the 2- or 3-position on a hydrocarbon chain. This was so
because when the halogen was in the 1-position it was exposed to more
collisions than when it was in the inner positions. In other words, the
halogen was shielded more by the molecule when it was in the inner positions. In summary, this paper is a very significant contribution to the
understanding of collision cross sections and shows a significant improvement in the precision of the diffusion data.

Wasik and McCulloh [21] have developed a variation on the basic GC
technique. They have solved the problem of finite injection volume by
allowing the solute to pass through a column C_1 and a detector D_1 directly
into the diffusion column C_2 and the detector D_2. They have developed
equations that describe the additional peak broadening in the second column
based on the peak profile after elution from the first column.

Using this method, Wasik and McCulloh have measured binary diffusion coefficients for argon, krypton, nitrogen, and oxygen diffusing into helium. The temperatures used ranged from 77 to 370°K (see Appendix Table I). Disadvantages to the method are few. The first column used was as long as the diffusion column, so that the peak profile would be Gaussian when it was "injected" into the diffusion column. This doubles the measurement time. Also there is some slight distortion of the peak caused by its passage through the first detector.

Hu and Kobayashi [40] have made extensive diffusion coefficient measurements by the GC method at pressures greater than 1 atm. They have used essentially the same theoretical treatment as that employed by Giddings. They have obtained a remarkable improvement in temperature control by using a liquid constant-temperature bath instead of the usual chromatographic oven. They measured the diffusion coefficient of $CH_3 T$ in CH_4 and CF_4 at 298°K and pressures of 2, 7, 20, 34, 47.5, and 61 atm, using a flow-through type of ionization chamber detector of their own design. They have also determined the diffusion coefficients of argon, methane, car carbon dioxide, and nitrogen diffusing into helium at temperatures of 248, 273, 298, and 323°K and at pressures of 10, 30, 50, and 60 atm, except no nitrogen-helium work was done at 298°K and the carbon dioxide-helium work was done at 10, 30, 40, and 50 atm. The precision was estimated to be within 1 to 3% (see Appendix Table I).

Nagata and Hasegawa [41] have used gas chromatography to determine diffusion coefficients for 20 binary systems. They used nitrogen and carbon dioxide as their carrier gases and investigated the following systems: helium, benzene, methyl acetate, carbon tetrachloride, chloroform, cyclo-hexane, ethyl formate, isopropanol, water, and acetone diffusing into nitrogen; and nitrogen, helium, benzene, methyl acetate, carbon tetra-chloride, cyclohexane, ethyl formate, isopropanol, and water diffusing into carbon dioxide. These systems were investigated at a variety of tempera-tures, ranging from 310.2 to 423.4°K (see Appendix Table I). Additionally, the results were compared with the diffusion coefficients calculated from four different predicting equations, including the method of Fuller, Schettler, and Giddings [31]. The latter method was the best overall, giving an average percent deviation of 5.1%.

Lozgachev and Kancheeva [42] have developed still another method based on frontal analysis. They claim to be able to determine mutual diffusion coefficients of an n-component mixture by first saturating a column with a mixture of n - 1 gases to be analyzed and then eluting the gases with a carrier. The diffusion coefficients are then determined from the shapes of the eluted fronts with the help of the equations derived by the authors. Although the method has merit and deserves further investigation, at present it suffers from rather poor precision, which for a three-component mixture was on the order of 20%.

Huang et al. [43] have investigated the effects of pressure and temperature on gaseous diffusion. They based their calculations on a modified version of our Eq. (6):

$$H = \frac{2D_{AB}}{\bar{U}} + \frac{r_0^2 \bar{U}}{24D_{AB}} + \frac{7\bar{U}r_0^4}{48R_0^2 D_{AB}} \tag{20}$$

where R_0 is the radius of the coil and the third term on the right-hand side accounts for the secondary flow phenomena due to column bending and the so-called race-track effect. Under their experimental conditions, where $r_0 = 0.25$ and $R_0 = 4.20$ cm,

$$D = \bar{U}/4[H \pm (H^2 - 0.0211)^{1/2} \tag{21}$$

Their quoted value of 0.0298 in the square-root term is in error. However, this will have a negligible effect on their results. Huang and co-workers noted that the measured value of the diffusion coefficient was velocity dependent, and they obtained their diffusion coefficients by plotting D_{AB} versus $\bar{U}$ and extrapolating to zero velocity. They reported diffusion coefficients for the gas-gas systems hydrogen-carbon dioxide and hydrogen-nitrogen; for several gas-liquid vapor systems (methanol, benzene, n-hexane, ethanol, cyclohexane, n-butanol, toluene, isooctane, oxylene, and sec-butanol diffusing into hydrogen), and for the gas-solid vapor systems hydrogen-naphthalene and hydrogen-camphor (see Table I).

Huang et al. presented an equation for estimating D_{AB} from known molar volumes V_A and V_B, the molecular weights M_A and M_B, the pressure P, the temperature T, and a correlation factor, A. The temperature dependence they found was $D \propto T^{1.75}$, which is in agreement with the results of other workers [31]. Curiously, over their pressure range of 750 to about 1700 mm Hg they found that the pressure dependence of D_{AB} was $D_{AB} \propto 1/P^{1.286}$. Most workers have assumed simple inverse pressure dependence. Their equation then is

$$D = \frac{AT^{1.75}}{(V_A^{1/3} + V_B^{1/3})P^{1.286}} \left(\frac{1}{M_A} + \frac{1}{M_B} \right)^{1/2} \tag{22}$$

The correlation factor A was determined to be 5.06. The average error in estimating D_{AB} by this method was 3.5%.

Liner and Weisman [44] used a tube furnace to make some high-temperature measurements of diffusion coefficients. They also determined the self-diffusion coefficient of helium using the ^{4}He-^{3}He system. The helium-argon and helium-nitrogen pairs were investigated at seven temperatures ranging from 303 to 806° K. The ^{4}He-^{3}He system was also investigated in the same temperature range. The maximum deviation from literature values was about 8%. The average deviation was estimated at 3 to 5% (see Appendix Table I).

Grushka and Maynard [45] have further demonstrated the usefulness of the GC technique by showing how it can be easily used in an instrumental analysis course to demonstrate both chromatographic theories and physico-chemical parameter measurement. Although the conditions of the analysis were probably the least well controlled of all the papers discussed here, the results still had a precision of about 5% (see Appendix Table I).

Grushka and Maynard [20] have also demonstrated how precise, and presumably accurate, the GC method may be. They determined the diffusion coefficients of seven octane isomers in helium in order to investigate the effect of molecular geometries. An extremely precise chromatographic system was built incorporating a fast-switching injection valve, precise temperature control (±0.1°C), and computer data reduction. The chromato-graphic peaks were digitized with a voltage-to-frequency converter. The diffusion coefficients were then calculated from the variances of Gaussian peaks, which were least-square fitted to the actual data. The least precise results had a relative standard deviation of about 1%. The overall relative standard deviation was about 0.34%.

Also, it was shown that there is a linear dependence of diffusion coefficient on critical volume: the smaller the critical volume, the larger the diffusion coefficient. Giddings's equation for estimating diffusion coefficients [30, 31] was modified to allow for estimation of isomers.

For the sake of completeness, we mention the paper by Arai, Saito, and Maeda [46]. Unfortunately a translation of this paper was not available in time for inclusion.

An important paper, which we have ignored to this point because it deals with dense gases, is the one by Balenovic, Myers, and Giddings [47]. These authors have done GC diffusion measurements at pressures of up to 1360 atm, where the density approaches that of a liquid. The method they used was the same in principle as the methods already discussed, but the equipment was modified because of the extremely high pressures involved. Because of the uniqueness of these measurements (more than an order of magnitude higher pressure than previously attempted by the GC method) their data are presented separately in Appendix Table II. This work indicates the versatility of the GC approach to diffusion measurements.

APPENDIX

TABLE I[a]

Diffusion Coefficients Measured by the Chromatographic Broadening Techniques

Trace solute	D_{AB}	Precision[b] (%)	Accuracy[c] (%)	T (°K)	P (atm)	Ref.
		Carrier gas H_2				
N_2	0.78	2.6	0	298	1	17
SF_6	0.57	—	29	292.6	1	15
CH_4	0.73	2.7	0	298	1	17
C_2H_6	0.54	1.9	1.9	298	1	17
C_3H_8	0.44	6.8	2.2	298	1	17
C_4H_{10}	0.40	3.8	—	298	1	17
$n-C_5H_{12}$	0.4895	1.5	—	353	1	36
	0.5324	3.9	—	373	1	36
	0.5830	4.3	—	393	1	36
	0.6300	0.06	—	423	1	36
	0.7425	0.47	—	453	1	36
$n-C_6H_{14}$	0.4990	0.94	—	353	1	36
	0.4740	1.5	10	373	1	36
	0.5310	0.38	—	303	1	36
	0.5923	0.30	—	423	1	36
	0.6520	0.0 (?)	—	453	1	36
Cyclohexane	0.5140	2.5	7.9	373	1	36
	0.5960	0.67	—	393	1	36
	0.6742	1.6	—	423	1	36
	0.7818	2.7	—	453	1	36
Benzene	0.5840	1.5	6.4	373	1	36
	0.6500	2.2	—	393	1	36
	0.7410	1.1	—	423	1	36

'race solute	D_{AB}	Precision[b] (%)	Accuracy[c] (%)	T ($^{\circ}$K)	P (atm)	Ref.
	Carrier gas H_2 — continued					
;enzene — continued	0.8220	2.2	—	453	1	36
	0.8940	0.94	—	483	1	36
'oluene	0.5834	—	—	373	1	36
	0.6170	7.7	—	393	1	36
	0.6724	0.74	—	423	1	36
	0.7440	1.5	—	453	1	36
	0.8197	4.9	—	483	1	36
'leOH	0.9370	1.0	—	353	1	36
	1.0200	4.6	—	373	1	36
	1.1420	4.2	—	393	1	36
	1.2483	6.8	—	423	1	36
:tOH	0.7200	1.4	—	353	1	36
	0.7820	1.3	—	373	1	36
	0.8420	0.61	—	393	1	36
	0.9460	2.8	—	423	1	36
	1.0770	4.9	—	453	1	36
-Butanol	0.6479	—	—	373	1	36
	0.7110	3.2	—	393	1	36
	0.7983	2.2	—	423	1	36
	0.9097	4.0	—	453	1	36
	1.0240	5.4	—	483	1	36
-Butanol	0.6290	—	—	373	1	36
	0.6760	1.6	—	393	1	36
	0.7850	3.9	—	423	1	36
	0.8730	3.9	—	453	1	36
	0.9690	0.63	—	483	1	36

Trace solute	D_{AB}	Precision[b] (%)	Accuracy[c] (%)	T (°K)	P (atm)	Ref.
		Carrier gas [4]He				
[3]He	1.88	4.8	3.3	303	1	44
	3.06	1.3	3.4	403	1	44
	4.39	1.8	2.1	500	1	44
	6.08	2.8	4.1	600	1	44
	7.56	1.9	0.3	698	1	44
	9.63	3.9	0.2	806	1	44
		Carrier gas He				
H_2	1.132	3.4	20	298	1	28
N_2	0.0725	1.4	—	77.2	1	21
	0.678	0.74	—	296	1	21
	0.687	0.74	0.15	298	1	28
	0.687	0.87	—	298	1	29
	0.750	0.93	5.2	303	1	44
	0.8171	0.61	—	321	1	21
	0.766	1.0	—	323	1	29
	0.798	2.3	3.6	324	1	44
	0.837	2.6	1.4	343	1	45
	0.9251	0.76	—	348	1	21
	0.893	0.56	—	353	1	29
	1.0300	0.58	—	370	1	21
	1.077	1.9	—	383	1	29
	1.13	1.8	2.7	403	1	44
	1.200	1.6	—	413	1	29
	1.289	1.1	—	443	1	29
	1.569	0.45	—	473	1	29
	1.650	1.3	—	498	1	29
	1.65	1.2	1.2	500	1	44

Trace solute	D_{AB}	Precision[b] (%)	Accuracy[c] (%)	T (°K)	P (atm)	Ref.
		Carrier gas He — continued				
N_2 — continued	2.20	0.91	4.1	600	1	44
	2.81	0.71	5.3	698	1	44
	3.75	2.4	1.4	806	1	44
	0.0522	—	—	248	9.97	40
	0.0177	—	—	248	29.9	40
	0.0107	—	—	248	49.8	40
	0.00889	—	—	248	59.8	40
	0.0607	—	—	273	9.97	40
	0.0206	—	—	273	29.9	40
	0.0124	—	—	273	49.8	40
	0.0105	—	—	273	59.8	40
	0.0820	—	—	323	9.97	40
	0.0272	—	—	323	29.9	40
	0.0166	—	—	323	49.8	40
	0.0140	—	—	323	59.8	40
O_2	0.718	1.3	—	298	1	28
	0.729	1.4	—	298	1	29
	0.7361	0.68	—	298	1	21
	0.8472	0.71	—	320	1	21
	0.809	0.87	—	323	1	29
	0.987	0.30	—	353	1	29
	1.041	0.77	—	365	1	21
	1.120	1.4	—	383	1	29
	1.245	1.1	—	413	1	29
	1.420	0.56	—	443	1	29
	1.595	1.6	—	473	1	29
	1.683	1.1	—	498	1	29

Trace solute	D_{AB}	Precision[b] (%)	Accuracy[c] (%)	T (°K)	P(atm)	Ref.
		Carrier gas He — continued				
Ar	0.0710	1.3	—	77.2	1	21
	0.729	1.0	0.55	296	1	28
	0.7335	0.55	—	298	1	21
	0.729	1.2	—	298	1	29
	0.784	1.0	1.8	303	1	44
	0.809	1.2	—	323	1	29
	0.847	1.8	1.9	324	1	44
	0.8890	0.67	—	334	1	21
	0.978	1.0	—	353	1	29
	0.9917	0.61	—	357	1	21
	1.122	1.2	—	383	1	29
	1.237	1.1	—	413	1	29
	1.22	3.3	3.3	402	1	44
	1.401	1.4	—	443	1	29
	1.612	0.87	—	473	1	29
	1.728	1.3	—	498	1	29
	1.75	1.1	4.6	500	1	44
	2.52	1.6	0.4	600	1	44
	3.05	4.6	6.9	698	1	44
	4.05	2.5	3.5	806	1	44
	0.0541	—	—	248	9.97	40
	0.0184	—	—	248	29.9	40
	0.0111	—	—	248	49.8	40
	0.00931	—	—	248	59.8	40
	0.0634	—	—	273	9.97	40
	0.0215	—	—	273	29.9	40
	0.0130	—	—	273	49.8	40

Trace solute	D_{AB}	Precision[b] (%)	Accuracy[c] (%)	T (°K)	P (atm)	Ref.
		Carrier gas He — continued				
Ar — continued	0.0109	—	—	273	59.8	40
	0.0742	—	—	298	9.97	40
	0.0249	—	—	298	29.9	40
	0.0152	—	—	298	49.8	40
	0.0127	—	—	298	59.8	40
	0.0851	—	—	323	9.97	40
	0.0288	—	—	323	29.9	40
	0.0175	—	—	323	49.8	40
	0.0145	—	—	323	59.8	40
Kr	0.6491	0.62	—	298	1	21
	0.7372	0.54	—	322	1	21
	0.813	0.62	—	341	1	21
	0.904	0.66	—	366	1	21
NH_3	0.923	0.76	—	297	0.84	28
SF_6	0.38	—	—	292.6	1	15
CO_2	0.612	0.49	—	298	1	29
	0.678	1.8	—	323	1	29
	0.800	1.6	—	353	1	29
	0.884	0.90	—	383	1	29
	1.040	1.1	—	413	1	29
	1.133	1.6	—	443	1	29
	1.279	1.5	—	473	1	29
	1.414	2.0	—	498	1	29
	0.0454	—	—	248	9.97	40
	0.0151	—	—	248	29.9	40
	0.0116	—	—	248	39.8	40
	0.00920	—	—	248	49.8	40

Trace solute	D_{AB}	Precision[b] (%)	Accuracy[c] (%)	T (°K)	P (atm)	Ref.
		Carrier gas He — continued				
CO_2 — continued	0.0525	—	—	273	9.97	40
	0.0177	—	—	273	29.9	40
	0.0133	—	—	273	39.8	40
	0.0107	—	—	273	49.8	40
	0.0616	—	—	298	9.97	40
	0.0206	—	—	298	29.9	40
	0.0156	—	—	298	39.8	40
	0.0127	—	—	298	49.8	40
	0.0701	—	—	323	9.97	40
	0.0236	—	—	323	29.9	40
	0.0177	—	—	323	39.8	40
	0.0141	—	—	323	49.8	40
CH_4	1.005	—	—	373	1	30
	1.007	0.70	0.20	373	1	20
	0.0501	—	—	248	9.97	40
	0.0169	—	—	248	29.9	40
	0.0103	—	—	248	49.8	40
	0.00872	—	—	248	59.8	40
	0.0588	—	—	273	9.97	40
	0.0198	—	—	273	29.9	40
	0.0119	—	—	273	49.8	40
	0.0101	—	—	273	59.8	40
	0.0681	—	—	298	9.97	40
	0.0229	—	—	298	29.9	40
	0.0139	—	—	298	49.8	40
	0.0117	—	—	298	59.8	40
	0.0781	—	—	323	9.97	40

Trace solute	D_{AB}	Precision[b] (%)	Accuracy[c] (%)	T (°K)	P (atm)	Ref.
	Carrier gas He — continued					
CH_4 — continued	0.0265	—	—	323	29.9	40
	0.0159	—	—	323	49.8	40
	0.0134	—	—	323	59.8	40
$n-C_4H_{10}$	0.364	0.27	—	298	1	35
	0.477	2.1	—	372.6	1	35
	0.634	0.95	—	423	1	35
	0.797	0.75	—	473	1	35
$n-C_5H_{12}$	0.288	0.35	—	298	1	35
	0.422	0.71	—	372.6	1	35
	0.565	1.2	—	423	1	35
	0.695	1.7	—	473	1	35
$n-C_6H_{14}$	0.273	1.8	—	298	1	35
	0.390	1.5	—	372.6	1	35
	0.574	—	—	417	1	30
	0.513	2.5	—	423	1	35
	0.629	1.9	—	473	1	35
$n-C_8H_{18}$	0.3161	1.0	—	373	1	20
3-Methylheptane	0.3334	0.27	—	373	1	20
2,4-Dimethylhexane	0.3340	0.24	—	373	1	20
3-Ethylhexane	0.3363	0.21	—	373	1	20
3-Ethyl-2-methyl- pentane	0.3398	0.12	—	373	1	20
2,3-Dimethylhexane	0.3420	0.18	—	373	1	20
2,2,4-Trimethyl- pentane	0.3455	0.32	—	373	1	20
Benzene	0.367	2.5	4.6	298	1	35
	0.498	3.6	—	372.6	1	35
	0.614	0.16	—	423	1	35

Trace solute	D_{AB}	Precision[b] (%)	Accuracy[c] (%)	T (°K)	P (atm)	Ref.
Carrier gas He — continued						
Benzene — continued	0.610	0.33	—	423	1	29
	0.662	0.15	—	443	1	29
	0.715	0.42	—	463	1	29
	0.778	2.1	—	473	1	35
	0.766	1.6	—	483	1	29
	0.815	1.5	—	503	1	29
	0.861	0.12	—	523	1	29
CH_3OH	1.032	2.1	—	423	1	29
	1.135	1.7	—	443	1	29
	1.218	1.7	—	463	1	29
	1.335	2.7	—	483	1	29
	1.389	1.1	—	503	1	29
	1.475	0.61	—	523	1	29
C_2H_5OH	0.496	—	0.40	298	1	35
	0.821	1.1	—	423	1	29
	0.862	1.2	—	443	1	29
	0.925	1.6	—	463	1	29
	0.997	3.1	—	483	1	29
	1.048	0.48	—	503	1	29
	1.173	0.34	—	523	1	29
1-Propanol	0.676	2.4	—	423	1	29
	0.711	0.98	—	443	1	29
	0.761	2.4	—	463	1	29
	0.829	0.60	—	483	1	29
	0.896	0.56	—	503	1	29
	0.959	0.10	—	523	1	29
2-Propanol	0.677	3.2	—	423	1	29

Trace solute	D_{AB}	Precision[b] (%)	Accuracy[c] (%)	T (°K)	P (atm)	Ref.
Carrier gas He — continued						
2-Propanol — continued	0.732	0.68	—	443	1	29
	0.784	0.77	—	463	1	29
	0.834	1.3	—	483	1	29
	0.882	0.68	—	503	1	29
	0.988	2.0	—	523	1	29
1-Butanol	0.587	1.9	—	423	1	29
	0.653	0.92	—	443	1	29
	0.689	0.87	—	463	1	29
	0.746	0.80	—	483	1	29
	0.792	1.1	—	503	1	29
	0.841	0.12	—	523	1	29
-Pentanol	0.507	0.99	—	423	1	29
	0.536	0.75	—	443	1	29
	0.578	2.2	—	463	1	29
	0.636	1.1	—	483	1	29
	0.666	0.75	—	503	1	29
	0.729	0.96	—	523	1	29
-Hexanol	0.469	1.5	—	423	1	29
	0.496	1.4	—	443	1	29
	0.531	0.19	—	463	1	29
	0.584	2.1	—	483	1	29
	0.631	0.63	—	503	1	29
	0.686	0.44	—	523	1	29
Ether	0.310	0.32	—	298	1	35
	0.460	2.2	—	372.6	1	35
	0.607	1.3	—	423	1	35
	0.745	3.9	—	473	1	35

Trace solute	D_{AB}	Precision[b] (%)	Accuracy[c] (%)	T (°K)	P (atm)	Ref.
		Carrier gas He — continued				
Acetone	0.411	3.4	—	298	1	35
	0.638	2.7	—	372.6	1	35
	0.754	1.9	—	423	1	35
	0.889	1.9	—	473	1	35
3-Pentanone	0.33	3.0	—	300	~1	19
CH_2F_2	0.874	3.4	—	430.8	1	22
F_2CH-CH_3	0.754	2.0	—	429.6	1	22
1-Fluorohexane	0.492	1.2	—	431.6	1	22
Fluorobenzene	0.566	1.4	—	429.7	1	22
C_6F_6	0.453	1.8	—	428.7	1	22
4-Fluorotoluene	0.508	1.2	—	431.6	1	22
CH_2Cl_2	0.750	1.2	—	427.5	1	22
$CHCl_3$	0.624	1.9	—	429.1	1	22
$ClCH_2-CH_2Cl$	0.683	0.88	—	427.1	1	22
1-Chloropropane	0.631	1.4	—	427.5	1	22
1-Chlorobutane	0.555	1.8	—	429.2	1	22
2-Chlorobutane	0.561	1.4	—	429.1	1	22
1-Chloropentane	0.518	0.77	—	428.2	1	22
Chlorobenzene	0.542	1.1	—	430.9	1	22
Dibromomethane	0.665	1.1	—	427.7	1	22
Bromoethane	0.740	1.5	—	427.7	1	22
1-Bromopropane	0.592	1.5	—	428.2	1	22
2-Bromopropane	0.606	2.0	—	428.0	1	22
1-Bromobutane	0.545	1.1	—	426.6	1	22
2-Bromobutane	0.553	2.4	—	427.2	1	22
1-Bromohexane	0.461	1.7	—	427.5	1	22
2-Bromohexane	0.470	2.6	—	427.9	1	22

Trace solute	D_{AB}	Precision[b] (%)	Accuracy[c] (%)	T (°K)	P (atm)	Ref.
		Carrier gas He — continued				
3-Bromohexane	0.469	0.85	—	428.5	1	22
Bromobenzene	0.543	1.8	—	427.1	1	22
2-Bromo-1-chloropropane	0.570	2.8	—	427.2	1	22
Iodomethane	0.783	2.0	—	431.2	1	22
Iodoethane	0.648	2.0	—	428.4	1	22
1-Iodopropane	0.579	1.2	—	430.0	1	22
2-Iodopropane	0.579	2.1	—	430.2	1	22
1-Iodobutane	0.524	1.3	—	428.1	1	22
2-Iodobutane	0.545	2.4	—	427.1	1	22
		Carrier gas N_2				
H_2	0.687	2.0	0.58	273	1	28
	0.7975	1.9	4.7	293	1	3
	0.876	—	3.6	324	1	16
He (?)[d]	0.605	3.6	—	293	1	37
He	0.717	0.84	1.8	298	1	28
	0.784	—	—	310.2	1	41
	0.812	—	—	315.2	1	41
	0.858	—	—	323.2	1	41
	0.867	—	—	330.2	1	41
	0.898	—	—	334.4	1	41
	0.927	—	—	340.2	1	41
	0.948	—	—	346.2	1	41
	0.978	—	—	353.9	1	41
	1.025	—	—	394.5	1	41
O_2	0.23	—	4.3	298	1	12
	0.258	—	6.6	324	1	16

Trace solute	D_{AB}	Precision[b] (%)	Accuracy[c] (%)	T ($^\circ$K)	P (atm)	Ref.
		Carrier gas N_2 — continued				
O_2 — continued	0.23	—	4.3	353	1	13
SF_6	0.086	—	—	292.6	1	15
H_2O	0.441	—	—	393.2	1	41
	0.464	—	—	408.2	1	41
	0.508	—	—	423.4	1	41
CO_2	0.163	0.52	2.4	298	1	28
	0.186	—	4.1	324	1	16
CCl_4	0.110	—	—	353	1	13
	0.113	—	—	363.7	1	41
	0.124	—	—	383.2	1	41
	0.134	—	—	403.2	1	41
	0.147	—	—	423.2	1	41
$CHCl_3$	0.135	—	—	361.0	1	41
	0.143	—	—	383.2	1	41
	0.161	—	—	403.2	1	41
	0.173	—	—	418.2	1	41
C_2H_6	0.14	18	—	298	1	17
C_3H_8	0.11	—	—	298	1	17
$n-C_4H_{10}$	<0.07	—	—	298	1	17
	0.0954	—	0.63	298	1	35
	0.100	—	1.5	302.4	1	34
$n-C_5H_{12}$	0.136	—	—	353	1	39
$n-C_6H_{14}$	0.111	1.4	—	353	1	32
	0.175	—	—	353	1	13
Cyclohexane	0.124	—	—	363.2	1	41
	0.134	—	—	383.2	1	41
	0.149	—	—	403.2	1	41

race solute	D_{AB}	Precision[b] (%)	Accuracy[c] (%)	T (°K)	P (atm)	Ref.
		Carrier gas N_2 — continued				
$_2H_4$	0.145	—	4.1	286.6	1	15
	0.033	—	6.1	290.9	4.4	15
	0.165	0.61	—	291	0.99	18
	0.16	—	—	291	1	11
	0.170	—	2.4	302.6	1	34
enzene	0.133	—	2.2	353	1	39
	0.129	—	—	364.2	1	41
	0.140	—	—	378.2	1	41
	0.154	—	—	393.4	1	41
	0.163	—	—	403.2	1	41
	0.165	—	—	423.2	1	41
IeOH	0.250	—	10	355	1	39
tOH	0.185	—	7.6	353	1	13
	0.177	—	2.8	355	1	39
-Propanol	0.159	—	—	362.9	1	41
	0.168	—	—	383.2	1	41
	0.146	—	3.5	357	1	39
cetone	0.140	—	—	343.1	1	41
	0.145	—	2.1	353	1	13
	0.135	—	5.2	353	1	39
	0.154	—	—	363.3	1	41
	0.170	—	—	383.2	1	41
-Pentanone	0.095	6.3	—	300	~1	19
ethyl acetate	0.160	—	12	353	1	13
	0.168	—	14	357	1	39
	0.171	—	—	363.5	1	41
	0.192	—	—	383.1	1	41

Trace solute	D_{AB}	Precision[b] (%)	Accuracy[c] (%)	T (°K)	P (atm)	Ref.
Carrier gas N_2 — continued
Methyl acetate	0.209	—	—	403.8	1	41
Ethyl formate	0.131	—	—	343.7	1	41
	0.143	—	—	363.2	1	41
	0.158	—	—	383.2	1	41
	0.168	—	—	403.2	1	41
Ethyl acetate	0.137	—	12	355	1	39
Carrier gas O_2
He	0.737	0.58	—	298	1	28
Carrier gas Ar
H_2	0.78	—	1.3	292.6	1	15
SF_6	0.055	—	—	292.6	1	15
$n-C_4H_{10}$	0.104	3.8	—	298	1	35
	0.139	2.2	—	372.6	1	35
	0.170	1.8	—	423	1	35
	0.209	2.9	—	473	1	35
$n-C_5H_{12}$	0.0890	0.90	—	298	1	35
	0.115	7.8	—	372.6	1	35
	0.149	2.0	—	423	1	35
	0.186	2.2	—	473	1	35
$n-C_6H_{14}$	0.0845	3.3	—	293	1	35
	0.107	0.93	—	372.6	1	35
	0.145	2.1	—	423	1	35
	0.174	2.9	—	473	1	35
Benzene	0.108	0.93	—	298	1	35
	0.142	1.4	—	272.6	1	35
	0.169	0.59	—	423	1	35
	0.212	1.4	—	473	1	35

Trace solute	D_{AB}	Precision[b] (%)	Accuracy[c] (%)	T (°K)	P (atm)	Ref.
		Carrier gas Ar — continued				
Acetone	0.115	2.6	—	298	1	35
	0.175	1.7	—	372.6	1	35
	0.213	0.94	—	423	1	35
	0.249	1.2	—	473	1	35
3-Pentanone	0.074	2.7	—	300	~1	19
Ether	0.0849	1.9	—	298	1	35
	0.116	2.6	—	372.6	1	35
	0.165	1.8	—	423	1	35
	0.203	3.9	—	473	1	35
		Carrier gas CO_2				
H_2	0.665	0.38	4.1	298	1	28
He	0.633	—	—	313.8	1	41
	0.668	—	—	324.3	1	41
	0.720	—	—	332.3	1	41
	0.746	—	—	342.3	1	41
	0.794	—	—	353.1	1	41
	0.816	—	—	364.5	1	41
N_2	0.181	1.4	8.4	298	1	28
	0.201	—	—	313.7	1	41
	0.209	—	—	323.2	1	41
	0.232	—	—	332.6	1	41
	0.238	—	—	342.6	1	41
	0.251	—	—	354.2	1	41
	0.269	—	—	365.1	1	41
H_2O	0.297	—	—	393.8	1	41
	0.311	—	—	408.5	1	41
	0.333	—	—	423.3	1	41

Trace solute	D_{AB}	Precision[b] (%)	Accuracy[c] (%)	T (°K)	P (atm)	Ref.
		Carrier gas CO_2 —continued				
CCl_4	0.085	—	—	363.3	1	41
	0.093	—	—	384.3	1	41
	0.100	—	—	403.1	1	41
	0.111	—	—	423.0	1	41
$CHCl_3$	0.110	—	—	363.3	1	41
	0.120	—	—	383.3	1	41
	0.129	—	—	403.8	1	41
Cyclohexane	0.098	—	—	363.1	1	41
	0.108	—	—	383.0	1	41
	0.114	—	—	403.4	1	41
	0.126	—	—	423.4	1	41
Benzene	0.105	—	—	363.6	1	41
	0.116	—	—	378.0	1	41
	0.122	—	—	393.7	1	41
	0.130	—	—	408.2	1	41
	0.150	—	—	422.8	1	41
2-Propanol	0.110	—	—	362.8	1	41
	0.120	—	—	383.5	1	41
	0.135	—	—	403.3	1	41
	0.145	—	—	418.1	1	41
Methyl acetate	0.114	—	—	363.2	1	41
	0.126	—	—	383.2	1	41
Ethyl formate	0.099	—	—	333.5	1	41
	0.105	—	—	348.3	1	41
	0.116	—	—	362.9	1	41

Trace solute	D_{AB}	Precision[b] (%)	Accuracy[c] (%)	T (°K)	P (atm)	Ref.
		Carrier gas CH_4				
CO_2 (?)[d]	0.17	18	5.6	298	1	17
CH_3T	0.106	—	—	298	2.04	40
	0.0319	—	—	298	6.78	40
	0.0100	—	—	298	20.4	40
	0.00598	—	—	298	33.9	40
	0.00429	—	—	298	47.5	40
	0.00323	—	—	298	61.1	40
		Carrier gas CF_4				
CH_3T	0.0688	—	—	298	2.05	40
	0.0204	—	—	298	6.78	40
	0.00643	—	—	298	20.4	40
	0.00373	—	—	298	33.9	40
	0.00240	—	—	298	47.5	40
	0.00177	—	—	298	61.1	40

[a] This table is a compilation of all diffusion coefficients done by the GC method through 1972. Also included are some of the authors' data that were in press at the time of the compilation [20]. We have tried to be comprehensive and apologize in advance to anyone we may have slighted.

[b] Where this is blank, the authors gave no details about their precision. Otherwise, precision has been identified as $P \equiv$ deviation/$D_{AB} \times 100\%$, where the deviation was given by the author cited. It must be emphasized that frequently the "precision" is statistically meaningless due to the methods some authors used to calculate the deviation of their data. Where replicate data were given, the deviation was defined as deviation = range/2, where the range was the difference between the high and low values given.

[c] Where this is blank, there were no literature data to which the GC value could be compared. Otherwise accuracy has been defined as

$$A = \frac{|D_{AB} - D_{lit}|}{D_{AB} \times 100\%}$$

[d] The author did not make clear which was the trace solute and which was the carrier

TABLE II

Diffusion Coefficients of Dense Gases at 298°K ($D_{AB} \times 10^3 cm^2/sec$)

Carrier gas	Trace solute	Pressure used (atm)						
		272	408	543	681	818	1 090	1360
He	N_2	2.65	1.78	1.32	1.09	0.94	0.74	0.72
He	Ar	2.80	1.85	1.37	1.13	0.95	0.75	0.65
He	CH_4	2.43	1.62	1.22	0.98	0.85	0.74	0.71
He	C_2H_6	1.77	1.22	0.94	0.77	0.68	0.60	0.58
He	C_3H_8	1.51	1.05	0.80	0.66	0.58	0.52	0.50
He	$n\text{-}C_4H_{10}$	1.30	0.91	0.70	0.58	0.51	0.46	0.44
He	CF_4	1.85	1.22	0.94	0.78	0.65	0.57	0.54
N_2	H_2	2.20	1.40	1.05	0.87	0.76	0.63	0.59
N_2	He	2.00	1.23	0.91	0.73	0.62	0.53	0.50
Ar	H_2	1.83	1.15	0.92	0.76	0.65	0.52	0.45
Ar	He	1.95	1.25	0.95	0.78	0.65	0.53	0.45

NOTE ADDED IN PROOF

Since the completion of this review in early 1973, several other papers appeared using this method for gaseous diffusion coefficients measurements. These papers are:

C. Vandensteendam and S. Piekarski, J. Chimie Phys., 70, 600 (1973).
T. Tsuda and D. Ishii, J. Chromatogr., 87, 554 (1973).
W. A. Wakeham and D. H. Slater, J. Phys. B: Atom Molec. Phys., 6, 886 (1974).
W. A. Wakeham and D. H. Slater, J. Phys. B: Atom Molec. Phys., 7, 297 (1974).
T. C. Chu, P. S. Chappelear and R. Kobayeshi, J. Chem. Eng. Data, 19, 299 (1974).
E. Grushka and P. Schnipelsky, J. Phys. Chem., 78, 1428 (1974).
E. Grushka and V. R. Maynard, Chem. Tech., 4, 560 (1974).
V. R. Choudhary, J. Chromatogr., 98, 491 (1974).

The last two papers are short reviews.

REFERENCES

1. E. A. Mason and T. R. Marrero, in Advances in Atomic and Molecular Physics, D. R. Bates and I. Esterman, eds., Academic Press, New York, 1958, p. 155.

2. T. R. Marrero and E. A. Mason, J. Phys. Chem. Ref. Data, 1, 3 (1972).

3. J. C. Giddings and S. L. Seager, J. Chem. Phys., 33 1579 (1960).

4. G. Taylor, Proc. Roy. Soc. (London), A219, 186 (1953).

5. G. Taylor, Proc. Roy. Soc. (London), A223, 446 (1954).

6. G. Taylor, Proc. Roy. Soc. (London), A225, 473 (1954).

7. R. Aris, Proc. Roy. Soc. (London), A235, 67 (1956).

8. J. C. Giddings, Dynamics of Chromatography, Dekker, New York, 1965.

9. M. J. E. Golay, in Gas Chromatography, D. H. Desty, ed., Butterworths, London, 1958, p. 36.

10. J. J. van Deemter, F. J. Zuiderweg, and A. Klinkenberg, Chem. Eng. Sci., 5, 271 (1956).

11. J. H. Knox and L. McLaren, Anal. Chem., 35, 449 (1963).

12. H. J. Arnikar, T. S. Rao, and K. H. Karmarkar, J. Chromatogr., 26, 30 (1967).

13. H. J. Arnikar, T. S. Rao, and K. H. Karmarkar, Int. J. Electronics, 22, 381 (1967).

14. A. Bournia, J. Coull, and G. Houghton, Proc. Roy. Soc. (London), A261, 227 (1961).

15. E. V. Evans and C. N. Kenney, Proc. Roy. Soc. (London), A284, 540 (1965).

16. J. Bohemen and J. H. Purnell, J. Chem. Soc., 360 (1961).

17. P. Fejes and L. Czaran, Hung. Acta. Chim., 29, 171 (1961).

18. J. H. Knox and L. McLaren, Anal. Chem., 36, 1477 (1964).

19. J. K. Barr and D. T. Sawyer, Anal. Chem., 36, 1753 (1964).

20. E. Grushka and V. R. Maynard, J. Phys. Chem., in press.

21. S. P. Wasik and K. E. McCulloh, J. Res. Natl. Bur. Stand. U.S., 73A, 207 (1969).

22. E. N. Fuller, K. Ensley, and J. C. Giddings, J. Phys. Chem., 73, 3679 (1969).

23. J. Loschmidt, Sitzber. Akad. Wiss. Wien, 61, 367 (1870).

24. J. Loschmidt, Sitzber. Akad. Wiss. Wien, 62, 468 (1870).

25. J. Stefan, Sitzber. Akad. Wiss. Wien, 68, 385 (1873).

26. E. P. Ney and F. C. Armstead, Phys. Rev., 71, 14 (1947).

27. A. A. Westenberg and R. E. Walker, J. Chem. Phys., 26, 1753 (1957).

28. J. C. Giddings and S. L. Seager, Ind. Eng. Chem. Fundam., 1, 277 (1962).

29. S. L. Seager, L. R. Geertson, and J. C. Giddings, J. Chem. Eng. Data, 8, 168 (1963).

30. E. N. Fuller and J. C. Giddings, J. Gas Chromatogr., 3, 222 (1965).

31. E. N. Fuller, P. D. Schettler, and J. C. Giddings, Ind. Eng. Chem., 58(5), 19 (1966).

32. J. F. K. Huber and G. van Vught, Ber. Bunsenges. Phys. Chem., 69, 821 (1965).

33. G. T. Chang, Ph. D. thesis, Rice University, Texas, 1966.

34. J. C. Giddings and K. L. Mallik, Ind. Eng. Chem., 59(4), 18 (1967).

35. G. L. Hargrove and D. T. Sawyer, Anal. Chem., 39, 244 (1967).

36. T. -C. Huang, S. -J. Sheng, and F. J. F. Yang, J. Chin. Chem. Soc. (Taipei), 15, 127 (1968).

37. A. A. Zhukhovitskii, S. N. Kim, and M. O. Burova, Zavod. Lab., 34, 144 (1968).

38. J. Anderson and K. I. Saddington, Chem. Soc. Suppl., 381 (1949).

39. H. J. Arnikar and H. M. Ghule, Int. J. Electronics, 26, 159 (1969).

40. A. T. -C. Hu and R. Kobayashi, J. Chem. Eng. Data, 15, 328 (1970).

41. I. Nagata and T. Hasegawa, J. Chem. Eng. Japan, 3, 143 (1970).

42. V. I. Lozgachev and O. A. Kancheeva, Russ. J. Phys. Chem., 46, 714 (1972).

43. T. -C. Huang, F. J. F. Yang, C. -J. Huang, and C. -H. Kuo, J. Chromatogr., 70, 13 (1972).

44. J. C. Liner and S. Weissman, J. Chem. Phys., 56, 2288 (1972).

45. E. Grushka and V. Maynard, J. Chem. Ed., 49, 565 (1972).

46. K. Arai, S. Saito, and S. Maeda, Kagaku Kogaku, 31, 25 (1967).

47. Z. Balenovic, M. N. Myers, and J. C. Giddings, J. Chem. Phys., 52, 915 (1970).

Chapter 5

GAS-CHROMATOGRAPHY ANALYSIS OF POLYCHLORINATED
BIPHENYLS AND OTHER NONPESTICIDE ORGANIC POLLUTANTS

Joseph Sherma

Department of Chemistry
Lafayette College
Easton, Pennsylvania

I. INTRODUCTION

The purpose of this chapter is to review the available methodology for the gas-chromatography (GC) analysis of certain organic chemicals whose residues have been found in various samples. Included will be analytical methods for polychlorinated biphenyls, chlorinated naphthalenes, phthalate esters, chlorodioxins, and a few other related types of compounds. Such topics as the sources, synthesis, occurrence, ecological and toxicological effects, industrial uses, metabolism, and chemistry of these compounds will not be covered in any detail.

Methods for the determination of the above-mentioned compounds are generally similar to those used for the analysis of pesticide residues. Pesticide-analysis methods have been covered in detail in a number of recent books and reviews [1-4]. A residue analysis usually consists of the following steps:

1. Extraction of the compounds of interest from the sample

2. Cleanup procedures to remove interfering substances; if liquid chromatography is used for cleanup, preliminary fractionation of the compounds to be analyzed usually results

3. Qualitative identification by means of gas chromatography, with confirmation by other techniques, such as thin-layer chromatography (TLC) or mass spectrometry (MS)

4. Quantitative analysis

Procedures for each of these steps are outlined in this chapter for each compound class.

II. POLYCHLORINATED BIPHENYLS

Polychlorinated biphenyls (PCBs) are a group of industrially important chemicals prepared by the chlorination of biphenyl and sold in the United States and Canada under the trade name Aroclor. Aroclors consist of complex mixtures containing isomers of chlorobiphenyls and/or chloroterphenyls (PCTs) with different chlorine contents. The range of biphenyl preparations are designated Aroclor 1221 (21 wt% Cl, average 1.15 Cl atoms per molecule, 192 MW) to Aroclor 1268 (68% Cl, 8.70 Cl atoms per molecule, 453 MW). Aroclor preparations with the first digits 25 and 44 are 75 and 60% blends, respectively, of chlorinated biphenyls and terphenyls, and preparations whose designations begin with 54 are chlorinated terphenyls. The PCBs are characterized by high thermal and chemical stability, low water solubility, and a low but finite vapor pressure. In addition to uses as dielectric fluids, plasticizers, heat-transfer fluids, specialized lubricants, and fire retardants, PCBs have been reported in formulations with organochlorine pesticides.

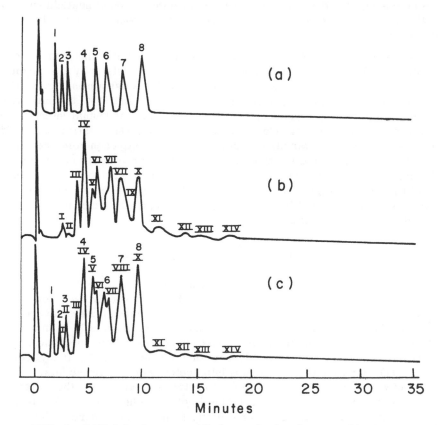

FIG. 1. PCB interference with the analysis of organochlorine pesticide residues in a borosillicate column, 6 ft x 1/8 in., packed with 4% SE-30 and 6% QF-1 on 60-80-mesh Chromosorb W. (a) Standard mixture of organochlorine pesticides. Peaks: 1, 0.08 ng of lindane; 2, 0.10 ng of heptachlor; 3, 0.10 ng of aldrin; 4, 0.14 ng of heptachlor epoxide; 5, 0.20 ng of DDE; 6, 0.20 ng of dieldrin; 7, 0.30 ng of DDD; 8, 0.50 ng of DDT. (b) Aroclor 1254, 5 ng (the 14 major peaks are numbered I to XIV). (c) Combination of the above organochlorine standard pesticide mixture and Aroclor 1254. Injector temperature 250° C; column temperature 200° C; tritium electron-capture detector base temperature 250° C; nitrogen flow rate, 20 to 30 ml/min. Redrawn with permission from Fig. 17, L. Fishbein, J. Chromatogr., 68, 364 (1972).

Residues of PCBs were first definitely identified in environmental samples in 1966, although there were earlier indications in the literature that these compounds were present in wildlife and air samples [5]. Since that time much research has been carried out, and PCB contamination has been found all over the world in such diverse samples as feeds, animal and fish tissue, human milk, wildlife, sewage, food products, water, and some organs of children. It has been suggested that PCBs may result from the action of sunlight on DDT and DDE [6].* General information on PCBs may be found in many published sources [7-14].

The complex nature of PCBs leads to complex problems when analyzing for residues. A large number of peaks result when analysis is by gas chromatography using the electron-capture detector (EC-GC). As shown in Fig. 1, GC elution times for PCB peaks (especially Aroclors 1254 and 1260) are often the same as those for many organochlorine pesticides, so that mutual interference may occur for PCB and pesticide determinations [14]. Quantitation has posed problems because of the variation in detector response to each component and the unavailability of absolute standards. Many of the procedures covered here are designed to deal with various PCB-pesticide mixtures. In some cases a separation is required prior to quantitation of both types of compound, whereas in other cases this may not be necessary, depending on the exact mixture and the levels present in the sample.

A. Extraction and Cleanup of Residues

Since PCBs and PCTs are fat-soluble, nonionic compounds, residues are completely, or at least partially [15], recovered through a widely used Food and Drug Administration (FDA) multiresidue method for the analysis of nonionic organochlorine pesticide residues in fatty and nonfatty foods [16]. This method involves extraction from fatty foods (or human and animal adipose tissues, etc.) by blending with petroleum ether (or hexane or 5% diethyl ether-hexane [17]), partition of the residues into acetonitrile to remove fats, addition of water, and back-extraction into petroleum ether. Further cleanup is obtained by chromatography on a Florisil column, eluting with a 6% ethyl ether-petroleum ether mixture. For nonfatty foods the procedure is the same except that the initial extraction is with acetonitrile or an acetonitrile-water mixture. If an activated (130°C) Florisil column (10 × 2.5 cm) is instead eluted with 200 ml hexane before 200 to 250 ml of a 20% mixture of diethyl ether and hexane, PCBs, PCTs, DDE, heptachlor, and aldrin are quantitatively (>92%) eluted in the first fraction, and such other chlorinated pesticides as DDT, DDD, dieldrin, lindane, hepatchlor epoxide, and endrin are eluted in the second fraction [18]. If this separation of PCBs from DDT is to be

---

*The chemical formulas and structures of pesticides mentioned throughout this chapter can be found in the Dictionary of Pesticides, published by Farm Chemicals, Meister Publishing Company, Willoughby, Ohio.

reproduced, a preliminary examination of the elution characteristics of Florisil prior to use is recommended [19]. A 10% ether-hexane mixture after hexane will yield similar results [20]. Silica columns have yielded elution patterns similar to these on Florisil [21, 22]. Samples of less than 1 g can be satis-factorily extracted by the unpublished Enos micromethod [4, 23].

The method recommended by the FDA for PCB-pesticide separation involves subjecting the 6% ether-petroleum ether Florisil eluate to chromato-graphy on a silicic acid-Celite (20:5, w/w) column in order to separate DDT and its analogs from some of the PCBs, including those with which they interfere. The silicic acid is standardized by adding enough water (usually ~3%) to provide maximum separation between p,p'-DDE and Aroclor 1254. Interfering PCBs and aldrin are eluted with 250 ml petroleum ether, followed by 200 ml acetonitrile-hexane-methylene chloride (1:19:80, v/v) to elute DDT compounds, most other organochlorine pesticides, and some other PCBs [24]. Contamination of the pesticide fraction with lower PCBs has been reported [15, 25, 26].

Other cleanup methods for Florisil eluates recommended by the FDA involve chemical reactions. Microscale alkaline hydrolysis [27-29] pro-vides not only additional cleanup for many sample types but also a method for confirming the identity of PCB residues. Although the stable PCB com-pounds will not be changed by alkali treatment, such compounds as DDT and p,p'-TDE will be converted to DDE and p,p'-MDE, respectively, which products would have different GC retention times. Oxidative treatment [30, 31] of cleaned-up sample extracts can be carried out with a chromic acid-acetic acid mixture to convert interfering DDE to 4,4'-dichlorobenzophenone prior to GC or TLC analysis of unchanged PCBs. Passage through a sulfuric acid-Celite cleanup column (Ref. 16, section 211.15b) does not alter PCBs [32].

A more recent alternative Florisil elution system has been designed by FDA workers [33] to provide improved cleanup for fats and oils, and recovery of pesticides of greater polarity compared with the original ether-petroleum ether system. The PCBs are eluted from the Florisil column along with many chlorinated pesticides with the first eluent in this new sequence: 20% methylene chloride in hexane. Silica gel has been further used for the difficult separation of PCBs from p,p'-DDE under carefully controled conditions [34, 35]. The purity of the hexanes used for these separations is critical in reproducing elution patterns [36]. Chlorinated pesticides are eluted from charcoal columns with 25% acetone in ethyl ether followed by elution of PCBs with benzene [37]. Cleanup of PCB residues in fish extracts has been based on gel-permeation chromatography on BioBeads S-X2 prior to separation of PCBs and pesticides on silicic acid [17, 38, 39]. Alkaline alumina (activity I) eluted with hexane has been used to frac-tionate PCBs in Aroclor 1254 formulations [40].

In addition to the blending technique, other extraction methods are as follows: PCBs in water samples at parts-per-trillion levels have been recovered by reversed-phase liquid-liquid partition between the water and a filter containing a 60-80-mesh Chromosorb W support coated with n-undecane and Carbowax 4000 [41]. Sorbed PCBs and pesticides are eluted with hexane. Residues may be adsorbed from water samples by porous polyurethane foam, with subsequent extraction with acetone and hexane [42]. Water samples have also been subjected to three-stage continuous liquid-liquid extraction with petroleum ether and benzene [22]. Extraction of PCBs from sewage solids was made using a mixture of hexane and isopropanol [43] and from lyophilized marine-diatom pellets with acetone [44]. Tissues of fish, crabs, oysters, and shrimp were extracted for 1 to 4 h with petroleum ether or hexane in a Soxhlet apparatus after mixing the sample with anhydrous sodium sulfate [20, 45, 46]. Chicken tissue was similarly extracted, followed by liquid-liquid partition with dimethylformamide and Florisil column cleanup [47]. Sediment samples were Soxhlet extracted for 4 h with 10% acetone-petroleum ether after drying at room temperature [45]. Blood samples were extracted with chloroform-methanol to recover PCTs [48]. A column-extraction procedure for fish, fish food, and mud samples combined extraction, filtration, and drying in one step [38, 49]. Aroclor 1254 (0.1 ppm) was isolated from fats and oils along with 0.005-ppm levels of organochlorine residues by distributing the fat or oil on a column of unactivated Florisil and eluting the residues by partitioning with 10% water in acetonitrile [50].

B. Gas-Chromatography Systems

Polychlorinated biphenyls are determined by gas chromatography using the electron-capture, microcoulometric, or electrolytic conductivity detectors. The construction, operation, and theory of these detectors have been previously described [1, 51, 52]. The electron-capture detector is the most sensitive but least selective of the three detectors for chlorine-containing compounds. The other two detectors are halogen specific, so that less cleanup of extracts may be required and confirmatory evidence for the identification of PCB residues will be obtained.

Recommended liquid phases and conditions for gas chromatography are generally the same as those for pesticide analyses [1, 3, 4]. As an example, Fig. 1 illustrates separations on a 6-ft × 1/8-in. o.d. column packed with a 4% SE-30/6% QF-1 mixed phase supported on acid-washed Chromosorb W and exhibiting about 2200 theoretical plates for DDT. The corresponding retention times of the PCBs and pesticides shown in Fig. 1 necessitate liquid-solid chromatography (e.g., on a Florisil column [18] as described in the preceding subsection) prior to chromatography of the eluates on the SE-30/QF-1 column, quantitation against an appropriate standard, and confirmation of identity (see next subsection). The Florisil technique [18] also separates PCBs from organophosphorus pesticides;

the flame-photometric GC detector [1, 53] in the phosphorus mode differentiates PCBs and organochlorine pesticides from phosphorus-containing pesticides.

Other liquid phases employed for PCB-pesticide analyses include 5% DEGS/2% H_3PO_4 [14], 10% DC-200 [21, 23, 24, 47, 54], and 15% QF-1/10% DC-200 (1:1) [24, 54] (Fig. 2). In all three cases PCBs were again found to be multicomponent mixtures with retention times in and beyond the range of organochlorine pesticides. With the last two columns and an electron-capture detector giving one-half full-scale recorder response for 1 ng heptachlor epoxide, 10 to 20 ng of various Aroclors was easily detected. With the latter column and a chloride microcoulometric detector, 0.6 μg Phenoclor DP-6 gave one-half scale response (1-mV recorder, 200-ohm microcoulometer range) [47].

Also used have been 5- to 9-ft packed columns at about 180 to 200°C containing SE-52, Apiezon-L, and XE-60 [40, 55-57]; OV-17/3% XE-60 and 12% DEGS [58, 59]; 3% QF-1 [32, 60-62]; Silicone/Epikote on Diatomite CQ [63]; 4% SE-30/2% QF-1 or 1.5% OV-17/1.95% QF-1 on Chromoport XXX [44]; 4% SF-96, 8% QF-1, or SF-96/QF-1 (3:1) on 100-200-mesh Gas-Chrom P [41, 64]; 5% DC-11 or 5% DC-11/15% QF-1 (1:1) on 60-80-mesh Chromosorb W [65]; 1.5% SE-30 on 60-80-mesh Chromosorb W [66]; 4% SE-30 on 100-120-mesh Chromosorb W (see Table I) [34, 67]; 3% OV-1 on 80-100-mesh Gas-Chrom Q [68]; 5% OV-17 on 60-80-mesh Gas-Chrom Q [69]; and 0.3% OV-7 on 80-100-mesh Corning 110 glass beads (155°C) [70]. In all cases PCBs produce multiple peaks with retention times similar to those for pesticides. For some applications long capillary or SCOT columns are required for the resolution of complex PCB and PCB-pesticide mixtures [17, 40, 71-73]. Separations of mono-, di-, and trichlorobiphenyls on polyphenyl thioether phases have been patented [74].

A study has been made [34] of retention times and electron-capture-detector responses of a series of standard PCBs and Aroclors 1254 and 1260 (which are of the type most often found in environmental samples). The effects of the number of chlorine substituents and their positions can be seen in Table I.

Albro and Fishbein [75] reported retention indices on six GC liquid phases for mono-, di-, and trichlorobiphenyls, and confirmed the additivity of half-retention index values in predicting retention indices. The liquid phases were 10% OV-101, 5% Versilube F-50, 6% Apiezon L, 10% OV-17, 10% OV-225, and 3% CHDMS on different supports in columns of varying lengths.

Methods for the GC analysis of phenolic PCB metabolites have been developed [76-78]. These may involve silylation or esterification reactions to block hydroxyl groups.

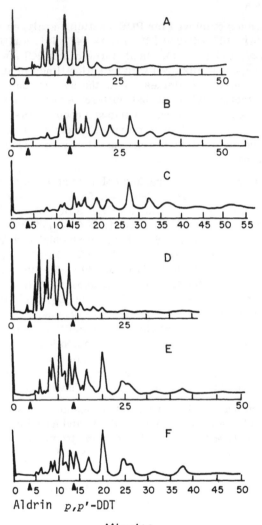

Aldrin *p,p'*-DDT

Minutes

Polychlorinated terphenyls are not eluted from 5% DEGS-2% H_3PO_4 or SE-30 GC columns at 200°C but can be determined on a 6-ft column of 3% OV-210 on 60-80-mesh Chromosorb W-AW at 200°C. Aroclor 5460 gives eight major peaks with retention times relative to decachlorobiphenyl of 1.78, 2.45, 3.10, 3.96, 4.83, 5.88, 7.25, and 9.13. The electron-capture-detector response based on the sum of the heights of peaks 3.10, 3.96, and 4.83 is 0.122 ng^{-1} relative to decachlorobiphenyl [9]. A 6-ft column packed with 3% Dexsil 300 on 100-120-mesh Gas Chrom Q at 310°C produced some 13 peaks eluting in 20 min for Aroclor 5460 with 10% methane-argon carrier gas at 40 ml/min. Minimum detection level was 2.5 ng using a pulsed [63]Ni electron-capture detector at 325°C [48].

C. Identification and Confirmation

Tentative identification of residues may be made by comparing retention times for all chromatographic peaks relative to a standard compound such as dieldrin with relative retention values of peaks from standard pesticides and commercial PCB formulations, usually on two or three GC columns of different polarity (e.g., DC-200, QF-1, DC-200/QF-1, DEGS) [29, 45]. Different Aroclors exhibit different GC patterns useful in distinguishing between them. These differences may include the number of major peaks and the peak-height ratios of certain peaks [14]. For confirmation of PCB and pesticide peaks, the position of elution from liquid-chromatography cleanup columns, derivatization and observation of any altered retention times [79], and the response of specific detectors are helpful. The Coulson chloride-selective detector has been modified for detection of chlorinated insecticides in the presence of PCBs [80]. Carbon-skeleton chromatography has been shown to differentiate between PCBs and DDT-type pesticides in samples of about 1 μg or less [81]. The carbon-skeleton chromatograms (flame-ionization detector) produced by PCBs and biphenyl are the same but very different from that of DDT. A mixture of cyclo-hexylbenzene and biphenyl results from Aroclor 1260 at a catalyst temperature of 300°C.

FIG. 2. Gas-chromatography separation of (A and D) 32 ng of Aroclor 1254, (B and E) 20 ng of Aroclor 1260, and (C and F) 20 ng of Aroclor 1262 on 10% DC-200 (A-C) and 1:1 15% QF-1/10% DC-200 (D-F). Columns 6 ft × 4 mm, 80-100-mesh Chromosorb W-HP support, nitrogen carrier flow 120 ml/min, column and electron-capture-detector temperature 200°C, injection-port temperature 225°C. Arrows show positions at which the pesticides aldrin and p,p'-DDT would elute under the given conditions. Redrawn with permission from Figs. 3 and 6, J. Armour, J. Chromatogr., 72, 275 (1972).

TABLE I

Retention Time and Electron-Capture-Detector Response of Chlorobiphenyls[a]

Compound	Relative retention time[b] (p,p'-DDE = 1.00)	Relative response per nanogram ± S.D. (p,p'-DDE = 1.00)
4-Chlorobiphenyl	0.17	0.0033 ± 0.00005
2-Chlorobiphenyl	0.11	0.0030 ± 0.00004
3-Chlorobiphenyl	0.14	0.0006 ± 0.00001
4,4'-Dichlorobiphenyl	0.30	0.0152 ± 0.0002
3,3'-Dichlorobiphenyl	0.25	0.0155 ± 0.0001
2,2'-Dichlorobiphenyl	0.15	0.0131 ± 0.0001
3,4-Dichlorobiphenyl	0.28	0.0388 ± 0.0002
2,4-Dichlorobiphenyl	0.19	0.0450 ± 0.0004
2,6-Dichlorobiphenyl	0.16	0.0815 ± 0.0049
2,4,4'-Trichlorobiphenyl	0.39	0.298 ± 0.010
2,4,6-Trichlorobiphenyl	0.24	0.276 ± 0.006
2,2',4,4'-Tetrachlorobiphenyl	0.51	0.206 ± 0.003
3,3',4,4'-Tetrachlorobiphenyl	0.96	0.770 ± 0.027
2,2',6,6'-Tetrachlorobiphenyl	0.33	0.0403 ± 0.0016

3,3',5,5'-Tetrachlorobiphenyl	0.70	0.625 ± 0.045
2,3,4,5-Tetrachlorobiphenyl	0.69	0.715 ± 0.010
2,3,5,6-Tetrachlorobiphenyl	0.51	0.505 ± 0.010
2,3,4,5,6-Pentachlorobiphenyl	1.00	1.30 ± 0.017
2,2',4,4',6,6'-Hexachlorobiphenyl	0.78	0.545 ± 0.029
3,3',4,4',5,5'-Hexachlorobiphenyl	2.98	1.15 ± 0.064
2,2',3,3',4,4',6,6'-Octachlorobiphenyl	2.62	1.58 ± 0.020
2,2',3,3',5,5',6,6'-Octachlorobiphenyl	2.45	1.53 ± 0.076
Decachlorobiphenyl	8.20	1.61 ± 0.086
Aroclor 1254	—[c]	0.910 ± 0.008
Aroclor 1260	—[d]	1.35 ± 0.010

[a] From Zitko, Hutzinger, and Safe [34].

[b] Glass column, 6 ft × 4 mm, containing 4% SE-30 on 100-200-mesh acid-washed Chromosorb, 200°C. Injection port and electron-capture detector, 210°C. Nitrogen carrier gas, 60 ml/min. Detector voltage 95 V dc, meter sensitivity 1×10^{-8} A.

[c] Fourteen peaks with relative retentions of 0.48 to 3.28 resulted.

[d] Sixteen peaks with relative retentions of 0.72 to 6.80 resulted.

TABLE II

Systems for Thin-Layer Chromatography of PCBs

Layer	Solvent	Detection	Remarks	Ref.
Aluminum oxide G incorporating $AgNO_3$	Benzene-hexane, 5:95	Ultraviolet light after spraying with phenoxyethanol-H_2O_2	DDE converted to DCBP by oxidative treatment; R_F PCBs = 0.91–0.94, DCBP = 0.30, other chlorinated pesticides = 0.48–0.88	30
MN–Silica gel G–HR/$AgNO_3$	n-Heptane; heptane-acetone, 98:2	Ultraviolet light	Two-dimensional development; PCBs separated from DDT and analogs	83
Silica gel	Hexane-acetone, 99:1		R_F PCBs = 0.8–0.1, organochlorine pesticides lower	55
Aluminum oxide G	Hexane-anhydrous diethyl ether, 40:0.8, or pure heptane	$AgNO_3$-2-phenoxy-ethanol-H_2O_2 spray followed by ultraviolet light	Aldrin and DDE overlap PCB spots when development is with heptane	65, 84, 85
Liquid paraffin (8%) on kieselguhr	Acetonitrile-acetone-methanol-water mixtures	1.7 g $AgNO_3$ in 96% ethanol spray followed by ultraviolet light	Reversed-phase TLC; PCBs appear as a number of distinct spots separated from DDE, DDT, and some other pesticides	86, 87

Waters Associates [82] obtained high-speed liquid chromatograms of PCBs by using Corasil/C_{18} packing with an acetonitrile-water mixture as the mobile solvent and an ultraviolet detector. However, this procedure has not yet been developed to a high degree of usefulness for residue analysis, and thin-layer chromatography [2] remains the most widely used chromatographic method for the confirmation of GC results. Some recommended systems for the thin-layer chromatography of PCBs are given in Table II. Thin-layer chromatography can also be used for quantitative analysis of PCB and pesticide residues, employing, for example, the fiber-optics densitometer designed by Beroza, Hill, and Norris [88] for pesticide determinations (now commercially available from Kontes, Vineland, N.J.).

Although ultraviolet, infrared [73, 89] and nuclear magnetic resonance [40, 90] spectroscopies have all been used on occasion to aid identification of PCB mixtures, mass spectrometry coupled with gas chromatography is undoubtedly the best procedure for confirming the presence of PCB residues and characterizing the molecular composition of PCB formulations [71, 91]. Coupling of the gas chromatograph and mass spectrometer is normally carried out in such a way that a chromatogram-type readout analogous to that produced by conventional GC detectors is produced. The usual source of the GC signal is a total-ion-current monitor located between the ion source and mass-analyzing magnet. The monitor intercepts a portion of the total ion beam and converts this current to a voltage output that is displayed on a potentiometric recorder [92]. Various interfaces between the gas chromatograph and the spectrometer have been used for residue work (see, for example, Refs. 93 and 94).

An early review of GC-MS analysis appeared in Chapter 6 of Volume 4 in this series, and a recent description of directly coupled GC-MS has been published [92]. An extensive review of MS applications to pesticide-residue analysis has been written by Biros [95]. Bonelli [96] reports that cleanup of PCB samples and separation from pesticides are not as important with GC-MS techniques as with conventional gas chromatography. Figure 3 shows that the mass spectrum of Aroclor 1254 is quite simple, although the preparation is a complex mixture. Detailed investigations of primary ion mass spectra and fragmentation patterns of synthesized individual PCB isomers have been made [97, 98]. It was concluded [99] that the primary ion spectra of different isomers are virtually undistinguishable with the result that the use of mass spectrometry for structural studies of PCBs is limited.

Mass spectrometry has been used to identify PCBs in human adipose tissue [91] after electron capture and microcoulometric GC indicated their presence. Programmed-temperature gas chromatography using a stainless-steel, 100-ft × 0.020-in. capillary column coated with OV-1 was applied to Aroclor standards and tissue extracts cleaned up on Florisil columns. Mass spectra were scanned over a range of m/e 5 to m/e 60 in 6 sec.

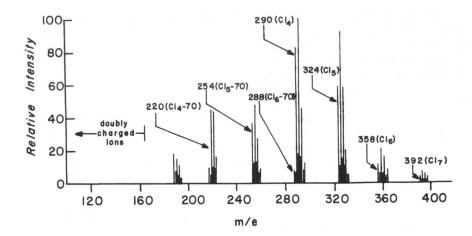

FIG. 3. Mass spectrum of Aroclor 1254. Redrawn with permission from Fig. 8, O. Hutzinger, S. Safe, and V. Zitko, Analabs Res. Notes, 12(2), 9 (1972).

Mass spectral data from a number of the total-ion-current monitor chromatographic peaks indicated the presence of PCBs and pesticides in the tissue samples after comparing samples with standards. Other GC-MS studies include identification of PCBs in bald eagles [100], identification of less than 10 ng of p,p'-DDE in the presence of Aroclor 1254 [101], identification of multiple residues of pesticides and PCBs in foods [25], studies of the occurrence and distribution of PCBs in the Rhine River [20], discrimination between PCBs and DDT [102], and identification of PCBs and DDT in mixtures and crude samples [103]. Stalling [104] has described the computer-assisted GC-MS examination of fish extracts for confirmation of PCB and pesticide residues.

D. Quantitation

Because of such factors as selective metabolism [72] and photochemical breakdown [105], GC peak patterns (both height and area ratios) obtained from biological and environmental samples seldom are exactly the same as the patterns from Aroclor standards. Accurate quantitation is therefore a difficult process, which has been attacked in different ways by different scientists. The FDA [24] and others [106] stipulate comparison of the total area of detector response for the residue to the total area of response obtained in the same way for a known weight of the commercial Aroclor standard with the most similar GC pattern. If the presence of more than one Aroclor is indicated, the residue should be quantitated using standards judged appropriate by the analyst for respective portions of the GC curve [15, 24]. Electron-capture, microcoulometric, and electrolytic conductivity chromatograms may be quantitated in this manner.

Other techniques for quantitation are based on the area of one or more selected peaks [20, 44, 63, 107, 108]; total peak height, or average or individual heights of selected peaks [23, 47, 61, 62, 108-111]; average electron-capture-detector response to biphenyls containing one to seven chlorine atoms [112]; determination of chlorine contents of different PCB peaks with the microcoulometric detector [47]; and perchlorination of all PCB compounds with $SbCl_5$ to a single derivative, namely, decachloro-biphenyl [37]. The latter procedure has been modified [113] to achieve a microscale (1 to 20 μg PCB residues) conversion of more than 90%, providing an easy method of quantitating and identifying total residues (without regard to composition) and increased detection sensitivity. Results obtained by this method (decachlorobiphenyl chromatographed on 1% OV-101) compared closely with those obtained using total area quantitation of unre-acted PCB after mathematical conversion to an equivalent amount of the Aroclor used as reference in the total area method. Perchlorination has also been used as a confirmatory test for chloroterphenyls found in the environment [114]. Reagents, conditions of formation, and properties of perchloro derivatives of biphenyl, terphenyl, naphthalene, dibenzofuran, dibenzodioxin, and DDE have been investigated [115], and analytical applications have been discussed [116]. A 3% OV-210 column at 200°C was used for gas chromatography of the fully chlorinated derivatives.

Changes in pesticide peak heights caused by chemical reactions such as nitration and saponification are valuable for quantitating PCBs in the presence of DDT-type compounds [61, 62]. Stalling [117] proposed mass spectro-metry (to determine the number of chlorine atoms per molecule) combined with programmed-temperature microcoulometric gas chromatography on SCOT columns as the way to achieve reliable quantitative analysis of PCB mixtures involving a large number of isomers. Bagley and Cromartie rely on semiquantitative thin-layer chromatography rather than gas chroma-tography for quantitation [30, 69].

The electron-capture response to different chlorobiphenyl compounds (which increases nonuniformly with chlorine content and varies according to substitution patterns) has been studied in detail [34], but standard PCBs and individual chlorinated biphenyl compounds have not been readily avail-able for purposes of quantitation, and commercial Aroclor preparations are most often used. It is therefore important to have relative retention times, response data, and reference chromatograms available for different Aroclors, and Armour [54] has obtained this information for Aroclors 1221, 1242, 1248, 1254, 1260, and 1262 on the widely used 10% DC-200 and 1:1 10% DC-200/15% QF-1 FDA columns (Fig. 2).

Procedures for preparing and using well-defined quantitative PCB standards were reported by Webb and McCall [118], and standards are available from the authors at the Southeast Environmental Research Laboratory, Athens, Georgia. The weight of PCB present in each standard GC peak was calculated from the empirical formula of the compound represented by the peak (determined by the GC-MS technique) and the absolute amount of chlorine represented by the peak (determined by electrolytic conductivity gas chromatography; response is linear with respect to the number of chlorine atoms present at a furnace temperature of $800^\circ C$). Polychlorinated biphenyls from environmental samples are quantitated by using one or more of the standard Aroclors, weight percent data for each peak present in the Aroclor standards, and some simple computation rules based on the principle that the total amount of PCB present is the sum of the amounts from all the individual peaks. The procedure is as follows:

To quantitate PCBs, chromatograph known amounts of the standards. Measure the area for each peak. Using the tables of data determine the response factor (ng PCB/cm^2) for each peak. Chromatograph the sample and measure the area of each peak. Multiply the area of each peak by the response factor for that peak. Add the nanograms of PCB found in each peak to obtain the total nanograms of PCB present. Samples containing one Aroclor or more than one Aroclor can be quantitated by comparison with appropriate standards. Hutzinger, Safe, and Zitko [119] have also synthesized individual PCB standards.

One international sample-exchange program produced results with a precision of $\pm 20\%$ (7), which is quite reasonable considering the lack of standardized quantitation procedures. The lack of individual chlorobiphenyl standards precluded any estimation of the "accuracy" of the results. An FDA interlaboratory study [11, 120] involving fish samples spiked with 6 ppm Aroclor 1254 or 1260 (and therefore involving no question in the choice of Aroclor quantitation standard) reported average results for PCBs 25% lower than actual. Quantitative recovery was reported for 7.5 ppm of Aroclor 1248 added to chicken fat.

The ubiquity [85, 121] of PCBs requires that great care be taken during analysis to avoid contamination of samples. All materials used in the analysis must be checked for the possible presence of PCBs [108].

Collins, Holmes, and Jackson [122] have reported a routine analytical procedure applicable to the monitoring of a variety of samples, including wildlife; it comprises methods already mentioned here. First PCB and pesticide residues are extracted from samples and cleaned up on a Florisil column; partial separation of PCBs from organochlorine residues is then made on a silica-gel column, p,p'-DDE in the PCB fraction is oxidized, and PCBs are determined by gas chromatography and/or thin-layer chromatography.

A convenient method for PCB quantitation is as follows: Determine the peak heights and retention times of all the peaks attributable to PCB on the chromatogram from the Apiezon L plus 0.15% Epikote 1001 column. Multiply each individual peak height by the corresponding retention time and sum all the products so obtained.   Divide this sum by the product of the peak height and retention time for 1 ng of p,p'-DDE when injected on the same column.   The result is a direct estimate of the weight (in nanograms) of the PCB injected on the column [122].

III. CHLORINATED NAPHTHALENES

$$Cl_x \underline{\hspace{1cm}} \langle\text{naphthalene}\rangle \underline{\hspace{1cm}} Cl_x$$

Chlorinated naphthalenes (Halowaxes) are industrial chemicals with many properties and uses similar to PCBs [123]. Residues of these compounds have been found in environmental and wildlife samples [62]. Like PCBs, chlorinated naphthalenes yield multipeak gas chromatograms, which would interfere with the analysis of some organochlorine pesticides.

A. Extraction and Cleanup

Halowax 1014 (a mixture of tetra-, penta-, and hexachloronaphthalenes) was added at the 0.5-ppm level to samples of fish, milk, and spinach and recovered to the extent of 67, 68, and 90%, respectively, by the FDA multiresidue procedure for organochlorine pesticides (acetonitrile-petroleum ether partitioning, Florisil cleanup) [16]. Halowaxes 1014 and 1099 (mixtures of trichloro- and tetrachloronaphthalene) were eluted from the Florisil column by 200 ml 6% ethyl ether-petroleum ether (along with DDT and many other chlorinated pesticides), while Halowax 1051 (octa-chloronaphthalene) was found in both fractions from successive 200-ml elutions with 6 and 15% ethyl ether-petroleum ether.

The silicic acid-Celite column chromatographic procedure developed for separating PCBs from chlorinated pesticides [24] was shown [124] to separate Halowaxes 1014, 1051, and 1099 as well. Each Halowax was eluted by 250 ml petroleum ether, along with any PCBs and aldrin present, while most of the common chlorinated pesticides were recovered in the acetonitrile-methylene chloride-hexane eluate. Figure 4 illustrates the silicic acid column separation of Halowax 1014 from p,p'-DDT, p,p'-TDE, and p,p'-DDE. Separation of chloronaphthalenes plus PCBs from organo-chlorine pesticides also occurs when a column of silica gel (2.5% moisture) is eluted with hexane [122], and chlorinated naphthalenes and PCBs are

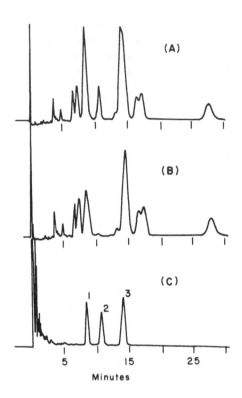

FIG. 4. Gas-liquid chromatography curves of brown-trout Florisil column eluate fortified with 2.5 ppm Halowax 1014, 0.3 ppm p,p'-DDT, and
0.2 ppm p,p'-TDE and containing 0.19 ppm p,p'-DDE residue, before and
after separation on silicic acid column. (A) Before separation. (B)
Petroleum ether eluate from silicic acid column containing chlorinated
naphthalene. (C) Polar eluate from silicic acid column containing (1)
p,p'-DDE, (2) p,p'-TDE, and (3) p,p'-DDT. A 10-mg sample was injected
for each curve. Conditions as in Fig. 2. Redrawn with permission from
Fig. 1, J. A. Armour and J. A. Burke, J. Chromatogr., 54, 175 (1971).

eluted together in fractions I and II (total 30 ml hexane) upon alumina-
silica chromatography as described by Holden and Marsden [9, 21, 108,
125]. Most of the DDE is eluted with the hexane eluent, while DDT,
dieldrin, TDE, and the remainder of the DDE are eluted with the second
eluent (fraction III), 10% ethyl ether in hexane.

B. Gas Chromatography

Figure 4 illustrates the conditions used for the electron-capture gas
chromatography of chlorinated naphthalenes. The results on columns

containing either 10% DC-200 or the mixed phase DC-200/QF-1 are similar
for a particular Halowax. Halowax 1099 gives 13 peaks ranging from 0.45
to 1.88 (relative to aldrin) on DC-200 and 10 peaks (0.49 to 1.96) on the
mixed phase. Halowax 1051 gives one peak (11.3 and 10.8, respectively)
on each column. A 10- to 25-ng quantity of the Halowaxes produces one-
half full-scale deflection under conditions giving the same response for 1
ng of heptachlor epoxide [124]. Columns containing 3% OV-101 (180°C)
and 1:1 3% OV-101/5% OV-210 (175°C) have been used to chromatograph
1.2 to 2.0 ng of Halowaxes 1013 and 1014 [126].

C. Confirmation of Identity

No method has been described as yet for the differential elution of chlorinated
naphthalenes and PCBs from liquid-solid adsorption columns. Likewise,
the TLC behavior of the two types of compounds is very similar [122, 127].
If chlorinated naphthalenes are present alone or in a concentration comparable to that of the PCBs with which they are mixed, characteristic electron-
capture GC peak patterns might allow their presence to be recognized [9],
although patterns are not as distinctive as those for PCB formulations [126].
Confirmation can be achieved by an oxidation method [128] in which chlorinated naphthalenes are destroyed by reaction with CrO_3, while the resistant
PCBs are not affected and will be again detected by the electron-capture
GC method.

Chlorinated naphthalenes can be confirmed in the presence of PCBs
and PCTs by ultraviolet spectrophotometry [67] if background impurities
permit. Chlorinated naphthalenes exhibit a maximum at 306 nm with an
$A_{1cm}^{1\%}$ of 329, while the latter compounds have negligible absorbance at
this wavelength [9]. Identification of chloronaphthalenes has also been
made by use of a Finnigan 150 GC-MS system [126].

IV. CHLORINATED DIBENZOFURANS

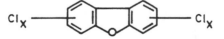

In a study by Vos et al. [129] chlorinated naphthalenes and the highly toxic
chlorinated dibenzofurans were identified in two commercial PCB formulations, and a toxicological evaluation was made of these impurities. Florisil
column chromatography, electron-capture and microcoulometric gas
chromatography on a 10% DC-200 column, and high-resolution mass
spectrometry were employed for the qualitative and quantitative determinations. Evidence was that some of the toxic effects of PCB preparations
may be due to the chlorodibenzofuran impurities.

TABLE III

Retention Time and Electron-Capture-Detector Response of
Chlorinated Dibenzofurans and Dibenzo-p-dioxins[a]

Compound	Retention time[b] (cm)	Response per nanogram	
		Height	Area
4% SE-30 Column			
Dichlorodibenzofuran	1.65	0.67	1.18
Trichlorodibenzofuran	3.65	0.63	3.36
Tetrachlorodibenzofuran	6.80		
Pentachlorodibenzofuran	13.5[c]		
Hexachlorodibenzofuran	27.6[c]		
Heptachlorodibenzofuran	55.0[c]		
Octachlorodibenzofuran	80		
2,7-Dichlorodibenzo-p-dioxin	1.90	0.28	0.49
Trichlorodibenzo-p-dioxin	3.64[c]		
2,3,7,8-Tetrachlorodibenzo-p-dioxin	7.35	0.043	0.25
Pentachlorodibenzo-p-dioxin	12.3[c]		
Hexachlorodibenzo-p-dioxin	22.8, 25.3	0.23	4.11
Heptachlorodibenzo-p-dioxin	43.6[c]		
Octachlorodibenzo-p-dioxin	83		
p,p'-DDE	3.95	6.03	16.3
3% OV-210 Column			
Octachlorodibenzofuran	4.05	0.22	
Hexachlorodibenzo-p-dioxin	1.40, 1.60	0.13	
Octachlorodibenzo-p-dioxin	4.30	0.23	
Decachlorobiphenyl	1.60	1.75	

No evidence of environmental contamination by dibenzofurans was found in a recent study [130]. Tetrachloro- and octachlorodibenzofurans were gas chromatographed on 4% SE-30 and 3% OV-210 (Table III) columns, respectively, after extraction, cleanup [46], and further liquid chromatography on activated ($130°C$) alumina columns eluted with 2 and 20% methylene chloride in hexane.

Perchlorination techniques have been used to assay samples of technical naphthalene and Halowax 1014 for the presence of dibenzofuran and chlorodibenzofuran. Alumina-column cleanup and gas chromatography on a 3% OV-210 column were employed [116].

Data obtained by the MS-GC method have shown that some commercial chlorophenols contained methoxy- and dimethoxypolychlorofurans, methoxy-polychloroethers, and polychlorohydroxybiphenyl. A 3% XE-60 column at $215°$ was employed [131].

V. CHLORINATED DIBENZODIOXINS

$$Cl_x - - Cl_x$$

These compounds (also known as "chick edema factor") first achieved notoriety when many of the mutagenic and teratogenic effects formerly attributed to the herbicide 2,4,5-T were shown [132] to be caused by a much more toxic [133] impurity, 2,3,7,8-tetrachlorodibenzo-p-dioxin ("tetra"; TCDD). Chlorodioxins may also be found in chlorophenols [134], which are widely used in agriculture and industry. Because of the high toxicities, analytical methods have been developed for detecting very low residue concentrations. Zitko [130] reported the absence of detectable quantities of chlorodioxins in the aquatic animals he studied.

---

[a]From Zitko and Choi [9].

[b]Retention times of Aroclor 1254 peaks on the SE-30 column were (1.95), 290, 3.45, (4.30), 5.15, 6.05, 7.20, (8.40), (10.1), (11.5), and (13.7). In parentheses are retention times of peaks not used for quantification.

[c]Estimated from plots of log (retention time) versus the number of chlorine atoms.

A. Extraction and Cleanup

Tetra is eluted together with PCBs and DDT and its analogs from Florisil
columns when 6% ether-petroleum ether [135] or 20% methylene chloride
in hexane [33] is the eluent. Octachlorodibenzo-p-dioxin is not recovered
from Florisil with either 6 or 15% ether-petroleum ether [16], but divides
between the final two methylene chloride-acetonitrile-hexane eluents in the
modified Florisil elution procedure [33]. These two dioxins are eluted
together with PCBs and polychlorinated naphthalenes by petroleum ether
from a silicic acid-Celite column [24].

Tetra, 2,3-dichloro-, and 2,3,7-trichlorobenzo-p-dioxins were separated from PCBs by column chromatography on activated (130°C) alumina
[135]. The PCBs were eluted with 1% methylene chloride-hexane, and
then chlorodioxins were eluted with 20% methylene chloride-hexane.
Alumina-column cleanup has been used prior to the GC analysis of the
chlorodioxins in fats [136, 137], chick tissues [138], aquatic animals
[130], and herbicide formulations [139]. Dioxins extracted from fat were
eluted in fractions I and II of the modified [9] Holden and Marsden [21]
silica-alumina cleanup.

A thin-layer cleanup procedure separates tetra and octachlorodibenzo-
p-dioxin from PCBs and organochlorine pesticides [140]. An aluminum
oxide layer is developed in unlined tanks with tetrachloroethylene-acetone
(1:4) for a distance of 7 cm from the bottom of the plate and, after drying,
with tetrachloroethylene-methanol-water (5:45:1) for 11 cm. Tetra- and
octachlorodioxins are located at the point reached by the first solvent,
while PCBs, DDT, DDE, DDD, and di- and trichlorodioxins move with the
second solvent front. The compounds can be eluted with benzene from
collected adsorbent for GC analysis.

Cleanup methods involving alumina or alumina-sulfuric acid [130, 131,
136, 139-141] were found inadequate for the sensitive (0.025 ppm) determination of chlorodioxins in the insecticide Gardona, and a sequential silica
gel (hexane eluent) and alumina (20% ether-hexane eluent) column chromatography followed by treatment of the resultant eluate with concentrated
sulfuric acid was used [142]. Steam distillation has been used to separate
tetra from 2,4,5-T prior to microcoulometric gas chromatography [143].

Extraction procedures for chlorodioxins are similar to those already
described [46] for the compounds covered earlier in this review. Fats,
oils, and fatty acids are dissolved in carbon tetrachloride; the solution is
then shaken with concentrated sulfuric acid and extracted with petroleum
ether [136, 137, 140]. Extractions from technical Gardona insecticide
formulations have been made using benzene [142], from 2,4,5-T formulations with diethyl ether [139], and from chlorophenols (dissolved in
alkali) with petroleum ether [131].

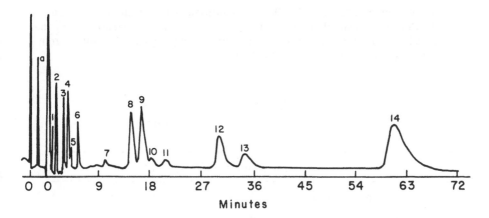

FIG. 5. Gas chromatogram of a mixture of chlorodibenzo-p-dioxins
(Cl_n-D): (1) 2-Cl-D; (2) 2,7-Cl_2-D; (3) 2,3,7-Cl_3-D; (4) 1,3,6,8-
Cl_4-D; (5) Cl_4-D isomer; (6) 2,3,7,8-Cl_4-D; (7) Cl_5-D isomer; (8-11)
Cl_6-D isomers; (12 and 13) Cl_7-D isomers; (14) Cl_8-D; (a) aldrin standard. Conditions: 7-ft × 4-mm i.d. glass coiled column packed with 3%
XE-60 on 80-100-mesh Gas-Chrom Q; column temperature 215° C; electron-
capture-detector voltage 80 V (tritium source); 1 × 10^{-9} A full scale;
nitrogen flow rate 55 ml/min. A 3-μl volume containing 4 to 140 ng of each
component was injected. Redrawn with permission from Fig. 1, A. Firestone et al., J. Assoc. Offic. Anal. Chem., 55, 88 (1972).

B. Gas Chromatography

Electron-capture [131, 139] and chloride-selective microcoulometric [143]
gas chromatography have been used to analyze chlorodioxins, with GC-MS
[144] confirmation of peaks. The following liquid phases have been applied:
3% OV-225 on 100-200-mesh Gas-Chroma Q, 200°C [142]; 5% OV-225 on
Chromosorb W, programmed temperature 230 to 260°C [144]; 2% OV-17
and 1% Hi-Eff 8 BP (cyclohexane dimethanol succinate), 200°C [139]; 10%
DC-200 on Gas-Chroma Q, 220°C (retention times were 1.55 and 2.72 for
tri- and tetrachlorodioxins, relative to aldrin) [143]; 3% XE-60 on 80-100-
mesh Gas Chrom Q, 175 to 215°C [131, 140] (Fig. 5); 3% OV-101 on
Chromosorb W-HP, 200°C [135]; and 4% SE-30 for tetra and 3% OV-210
for hexa-, hepta-, and octachlorodioxins (Table III) [130]. If the retention
times of dioxins and PCBs are the same, as is the case for tetra and the
common Aroclors on DC-200 and QF-1/DC-200 liquid phases [135], the
alumina-column LC separation described above is recommended.

Perchlorination has been used for a screening method to detect chlorodioxins at 0.05-ppm levels in the presence of PCBs, chloronaphthalenes,

and chlorodibenzofurans [145]. Octachlorodibenzo-p-dioxin is produced
from the dioxins, cleaned up on an alumina column, and separated from the
other perchlorinated derivatives by chromatography on a 3% XE-60 column
at 210°C. The results are shown in Table IV.

TABLE IV

Separation of Octachlorodibenzo-p-dioxin

	Retention time[a]	Quantity injected (ng)[b]
Octachlorodibenzo-p-dioxin	1.0	4.0
Octachlorodibenzofuran	0.90	3.5
Octachloronaphthalene	0.33	1.5
Decachlorobiphenyl	0.31	1.0

[a]Relative to octachlorodioxin, which eluted in 30 min on a 6-ft column
with 60-ml/min nitrogen carrier-gas flow.

[b]Required for one-half full-scale deflection with an electron-capture-
detector electrometer setting of 8×10^{-10}.

Because of their extreme toxicity, great care must be exercised in
handling chlorodioxin standards. A common standard for hexa-, hepta-,
and octachlorodibenzo-p-dioxin is 1.5% reference toxic fat in vegetable
oil available from the Division of Pesticide Chemistry and Toxicology, FDA,
Washington, D.C., 20204. A series of 13 chlorinated dioxins, containing
one to eight chlorine atoms, was prepared and characterized for use as
standards in analytical methods [146].

VI. PHTHALATE ESTERS

Esters of phthalic acid $[(C_6H_4(COOH)_2]$ are important industrial plasticizers
that have been found widely distributed in biological, food, and environmental samples [147-150]. Phthalate esters are in most cases high-boiling-
point liquids with very low vapor pressures. The most used phthalates
have been di-2-ethylhexyl (DEHP), diisodecyl, diisooctyl, n-octyl-n-decyl,
dibutyl, and diethyl.

Since these compounds are not chlorinated as are the other compound types covered in this chapter, GC analytical procedures are not so sensitive. Although the acute toxicity of the esters in quite low, really little is known about the details of their metabolism and toxicological effects. On the basis of presently available toxicological evidence and knowledge of their biological effects, Zitko [151, 152] has suggested that the phthalates are very likely present in the environment at "no effect" levels and the analytical detection limits may be sufficiently low. As new information is obtained, more sensitive analytical procedures might be required. Papers delineating what is known about the properties, utility, toxicity, metabolism, stability, and environmental sources of phthalate esters have been recently published [151, 153].

A. Extraction and Cleanup

Zitko [151, 152] has described a method for determining phthalates at 5- to 19-μg/g lipid levels in biological samples and commercial fish food which can be incorporated into procedures for chlorinated hydrocarbons. Phthalates were extracted with hexane and partially separated from lipids by chromatography on a 45 × 0.7-cm (2 g) alumina column deactivated with 5% water [21]. Elution was with 20 ml hexane, 20 ml 2% diethyl ether in hexane, and finally 20 ml 10% ether in hexane. The elution sequence of the various phthalates (and consequently their separation from PCBs and pesticides) depends on the lipid loading of the column. The fractions are evaporated to dryness, taken up in carbon disulfide (0.2 to 0.6 ml), and examined by gas chromatography with flame-ionization detection. An additional cleanup of alumina-column fractions containing high levels of lipids is required for TLC confirmation. This was achieved by partition between hexane and dimethylformamide. Other recommended extractants for phthalates are diethyl ether [154-156], methylene chloride [157], petroleum ether [158], and boiling methanol [159]. Phthalates are recovered, at least in part, from a Florisil column in the 15% ether-petroleum ether eluate [117, 147].

Phthalates and metabolic products were successfully extracted from fish by a variation of the column extraction procedure of Hesselberg and Johnson [49]. Tissue was blended with Na_2SO_4 and packed into a column that was then extracted by percolation with 1% H_3PO_4 in acetone. Unconjugated diesters, monoesters, and phthalic acid were separated from monoalkyl ester and phthalic acid conjugates by extraction procedures, the latter were enzymatically cleaved, polar products were derivatized with diazomethane, and coextracted fish lipids were removed from sample extracts by Florisil or gel-permeation chromatography prior to gas chromatography [160].

Dioctyl phthalate was isolated from milk by dialysis, freeze-drying, and extraction with petroleum ether. Cleanup was by column chromatography on alumina and thin-layer chromatography on silica gel [149]. Tissues were freeze-dried, and di-2-ethylhexyl phthalate was extracted with 20 volumes of chloroform-methanol (2:1); subsequent cleanup was by treatment with silicic acid prior to gas chromatography on a 3% SE-30 column at 225°C [161]. This compound is eluted in the 4% ether in hexane fraction from a silicic acid column [148].

Analytical methods for phthalate plasticizers have also been based on measuring the alcohol moiety. Dicyclohexyl and dibutyl phthalates were extracted from cheese and lard by methylene chloride. Following saponification, butanol and hexanol were isolated by steam distillation and determined by gas chromatography [162]. Di-2-ethylhexyl phthalate was analyzed similarly in milk [163].

Metabolites appearing in the urine of rats feed DEHP were isolated by extraction of the urine, after acidification, with diethyl ether; thin-layer chromatography of free metabolites and gas chromatography after treatment with diazomethane followed [164].

B. Gas Chromatography

Gas chromatography was carried out with a flame-ionization detector using a stainless-steel column (6 ft × 1/4 in.) containing 3% SE-30 on Anakrom ABS at 170 or 240°C (depending on the mixture), a nitrogen carrier-gas flow of 80 ml/min, an injection-port temperature of 250°C, and a detector temperature of 270°C [152]. Columns containing 1% polyvinylpyrrolidone and 1% Carbowax 20M on DMCS Chromosorb G [165]; 5% SE-30 on Chromosorb W (225°C) [148]; 2:1 SE-52/XE-60 (200°C), Apiezon L (200°C), and DEGS (180°C) [166]; and silicone grease (206°C) [167] have also been used for good separations of various phthalates. Nickel-63 electron-capture detection was used in conjunction with a column of 0.3% OV-7 on 80-100-mesh Corning GLC-110 glass beads at 155°C. Retention times relative to aldrin were 1.24 and 14.9 for di-n-butyl and diethylhexyl phthalates, respectively. Detector responses were 0.12 and 0.09 relative to DDT, with minimum detection limits of 0.1 to 0.5 μg/g of fish samples [147]. Figure 6 shows the separation of five esters on a lightly loaded column of OV-11 [160].

Phthalate metabolites were chromatographed on columns of 10% OV-3 at 180°C for determination of retention indices and 3% OV-1 for introduction into a mass spectrometer [164].

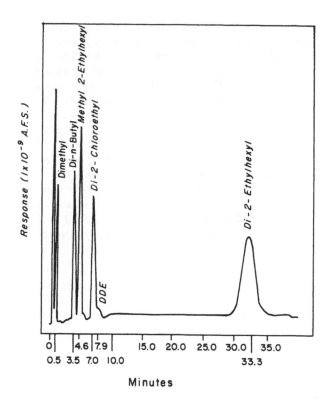

FIG. 6. Gas-liquid chromatogram of phthalate esters. The GLC
column, 183 cm x 2 mm i.d., was packed with OV-11 [0.3%, w/w, coated
on Corning GLC-110 (80-100 mesh) glass beads]: dimethyl phthalate, 3 ng;
di-n-butyl phthalate, 6 ng; methyl 2-ethylhexyl phthalate, 15 ng; di-2-
chloroethyl phthalate, 2 ng; di-2-ethylhexyl phthalate, 15 ng. Column
temperature 182°C; nitrogen carrier flow 30 ml/min. Redrawn with
permission from Fig. 4, D. L. Stalling, J. W. Hogan, and J. L. Johnson,
Environmental Health Perspectives, January 1973, p. 159.

C. Confirmation of Identity

Phthalates can be confirmed by measurement of flurorescence in concentrated sulfuric acid with excitation maxima at 270 and 308 nm and emission
maxima at 360 and 380 nm, respectively, for the ortho and para isomers.
No fluorescence was observed for m-phthalates [152]. As little as 0.06
μg/ml of phthalate (expressed as di-2-ethylhexyl phthalate) was detectable
in up to 100 μg/ml lipids.

Thin-layer chromatography can be carried out on silica-gel layers containing a phosphor, with chloroform as the mobile phase. Phthalate spots are detected directly under 245-nm ultraviolet light, by spraying with resorcinol in acidified ethanol, or by noting the yellow fluorescence on irradiation at 350 nm [152]. Other silica-gel systems include xylene-ethyl acetate-hexane (9:1:1) as solvent, with detection by spraying with 4 N sulfuric acid-20% ethanolic recorcinol (1:1) [154]; methylene chloride solvent, with resorcinol and methyl red as detecting reagents [168]; and ethyl acetate-cyclohexane (1:9) solvent [169], with Draggendorff reagent or anisaldehyde for detection [170]. Fuchsine dyes have been used to detect phthalates on thin-layer plates [171]. Chrom AR-1000 sheets and Eastman Chromagram Silica sheets developed with petroleum ether-diethyl ether-acetic acid (80:20:2-5, v/v) also provide chromatographic character-ization [160]. Other TLC systems have been described by Fishbein and Albro [153]. The high-speed liquid chromatography of phthalates on solid core supports was described by Majors [172].

Infrared spectroscopy and mass spectrometry are additional methods for identifying phthalates [117, 149, 160]. A catalog of reference infrared spectra has been published [173]. Both infrared and NMR spectroscopy as well as mass spectrometry were used to characterize DEHP urinary metabolites as ω- and (ω-1) oxidation products [164].

ACKNOWLEDGMENTS

The author expresses deep appreciation to the investigators who answered his requests for reprints and information on the topics covered in this review. These include A. V. Holden, D. L. Stalling, V. Zitko, O. Hutzinger, G. Westöö, S. Jensen, D. Firestone, B. M. McMahon, and G. C. Yang.

REFERENCES

1. G. Zweig and J. Sherma, eds., Analytical Methods for Pesticides and Plant Growth Regulators, Vol. VI, Gas Chromatographic Analysis, Academic Press, New York, 1972.

2. J. Sherma and G. Zweig, eds., Analytical Methods for Pesticides and Plant Growth Regulators, Vol. VII, Thin-Layer and Liquid Chromatography and Analysis of Pesticides of International Importance, Academic Press, New York, 1973.

3. J. Sherma and G. Zweig, in Chromatography, 3rd ed., E. Heftmann, ed., Reinhold, New York, in press.

4. J. Sherma, Critical Reviews in Analytical Chemistry, Vol. 3, August 1973, p. 299.

5. S. Jensen, in Proceedings of the PCB Conference, Stockholm, Sweden, September, 1970, National Swedish Environment Protection Board, Research Secretariat, December 1970, pp. 7-17.

6. R. W. Moilanen, Chem. Eng. News, April 23, 1973, p. 21.

7. O. Hutzinger, S. Safe, and V. Zitko, Analabs Res. Notes, 12, July 1972.

8. R. Edwards, Chem. Ind. (London), 1340 (1971).

9. V. Zitko and P. M. K. Choi, Technical Report No. 272, Fisheries Research Board of Canada, St. Andrews, New Brunswick, 1971.

10. N. S. Platonow, P. W. Saschenbrecker, and H. S. Funnell, Can. Vet. J., 12, 115 (1971).

11. D. L. Stalling and F. L. Mayer, Jr., Environmental Health Perspectives, April 1972, p. 159.

12. D. B. Peakall and J. L. Lincer, Bio Science, 20, 958, 1970.

13. Interdepartmental Task Force on PCBs, National Technical Information Service, Com-72-10419, May 1972.

14. L. Fishbein, J. Chromatogr., 68, 345 (1972).

15. J. A. Burke, J. Assoc. Offic. Anal. Chem., 55, 284 (1972).

16. Pesticide Analytical Manual, Vol. I, Sections 211 and 212, U.S. Department of Health, Education, and Welfare, Food and Drug Administration, Rockville, Md., first issued in 1967, with yearly revisions.

17. D. L. Stalling, paper presented at Meeting on Analysis of PCBs, September 17, 1971, sponsored by Working Group on Pesticides Monitoring Panel.

18. L. M. Reynolds, Bull. Environ. Contam. Toxicol., 4, 128 (1969); Residue Rev., 34, 27 (1971).

19. A. Bevenue and J. N. Ogata, J. Chromatogr., 50, 142 (1970).

20. J. H. Koeman, M. C. Tennoever de Brauw, and R. H. de Vos, Nature, 221, 1126 (1969).

21. A. V. Holden and K. Marsden, J. Chromatogr., 44, 481 (1969).

22. V. Leoni, J. Chromatogr., 62, 63 (1971).

23. D. J. Hansen, P. R. Parrish, L. I. Lowe, A. J. Wilson, Jr., and P. D. Wilson, Bull. Environ. Contam. Toxicol., 6, 113 (1971).

24. J. Armour and J. A. Burke, J. Assoc. Offic. Anal. Chem., 53, 761 (1970); Pesticide Analytical Manual, Vol. I, Section 251, January 1, 1972.

25. S. W. Bellman and T. L. Barry, J. Assoc. Offic. Anal. Chem. , 54, 499 (1971).

26. H. T. Masumoto, J. Assoc. Offic. Anal. Chem., 55, 1092 (1972).

27. S. J. V. Young and J. A. Burke, Bull. Environ. Contam. Toxicol. , 7, 160 (1972).

28. R. T. Krause, J. Assoc. Offic. Anal. Chem. , 55, 1042 (1972).

29. M. J. deFaubert Mauder, H. Egan, E. W. Godly, E. W. Hammond, J. Roburn, and J. Thomson, Analyst, 89, 168 (1964).

30. B. M. Mulhern, E. Cromartie, W. L. Reichel, and A. A. Belisle, J. Assoc. Offic. Anal. Chem. , 54, 548 (1970).

31. J. R. W. Miles, J. Assoc. Offic. Anal. Chem. , 55, 1039 (1972).

32. T. T. Schmidt, R. W. Risebrough, and F. Gress, Bull. Environ. Contam. Toxicol. , 6, 235 (1971).

33. P. A. Mills, B. A. Bong, L. R. Kamps, and J. A. Burke, J. Assoc. Offic. Anal. Chem. , 55, 39 (1972)   Pesticide Analytical Manual, Vol. I, Section 252, September 1, 1972.

34. V. Zitko, O. Hutzinger and S. Safe, Bull. Environ. Contam. Toxicol. , 6, 160 (1971).

35. D. Snyder and R. Reinert, Bull. Environ. Contam. Toxicol. , 6, 385 (1971).

36. V. Zitko, J. Chromatogr. , 59, 444 (1971).

37. O. W. Berg, P. L. Diosady, and G. A. V. Rees, Bull. Environ. Contam. Toxicol. , 7, 338 (1972).

38. D. L. Stalling, R. C. Tindle, and J. L. Johnson, J. Assoc. Offic. Anal. Chem. , 55, 32 (1972).

39. R. C. Tindle and D. L. Stalling, Anal. Chem. , 44, 1768 (1972).

40. D. Sissons and D. Welti, J. Chromatogr. , 60, 15 (1971).

41. B. Ahling and S. Jensen, Anal. Chem. , 42, 1483 (1970).

42. H. D. Gesser, A. Chow, F. C. Davis, J. F. Uthe, and J. Reinke, Anal. Letters, 4, 883 (1971).

43. A. V. Holden, Nature, 228, 1220 (1970).

44. J. E. Keil, L. E. Preister, and S. H. Sandifer, Bull. Environ. Contam. Toxicol. , 6, 156 (1971).

45. T. W. Duke, J. I. Lowe, and A. J. Wilson, Jr. , Bull. Environ. Contam. Toxicol. , 5, 171 (1970).

46. V. Zitko, Bull. Environ. Contam. Toxicol., 6, 464 (1971).

47. J. G. deVos and J. H. Koeman, Toxicol. Appl. Pharmacol., 17, 656 (1970).

48. R. F. Addison, G. L. Fletcher, S. Ray, and J. Doane, Bull. Environ. Contam. Toxicol., 8, 52 (1972).

49. R. J. Hesselberg and J. L. Johnson, Bull. Environ. Contam. Toxicol., 7, 115 (1972).

50. M. L. Porter and J. A. Burke, J. Assoc. Offic. Anal. Chem., 56, 733 (1973).

51. W. E. Westlake, in Advances in Chemistry Series 104, American Chemical Society, Washington, D.C., 1971, Chapter 5.

52. W. A. Aue and S. Kapila, J. Chromatogr. Sci., 11, 255 (1973).

53. W. A. Aue, in Advances in Chemistry Series 104, American Chemical Society, Washington, D.C., 1971, Chapter 4.

54. J. A. Armour, J. Chromatogr., 72, 275 (1972).

55. D. C. Holmes, J. H. Simmons, and J. O'G. Tatton, Nature, 216, 227 (1967).

56. J. H. Simmons and J. O'G. Tatton, J. Chromatogr., 27, 253 (1967).

57. D. T. Williams and B. J. Blanchfield, J. Assoc. Offic. Anal. Chem., 54, 1429 (1971).

58. W. L. Reichel, T. G. Lamont, and E. Cromartie, Bull. Environ. Contam. Toxicol., 4, 24 (1969).

59. W. C. Krantz, B. M. Mulhern, G. E. Bagley, A. Sprant, F. J. Ligas, and W. B. Robertson, Jr., Pestic. Monit. J., 4, 136 (1970).

60. R. W. Risebrough, P. Reiche, D. B. Peakall, S. G. Herman, and M. N. Kirven, Nature, 220, 1098 (1968).

61. R. W. Risebrough, in Chemical Fallout, M. W. Miller and G. G. Berg, eds., Thomas, Springfield, Ill., 1969, p. 5

62. R. W. Risebrough, P. Reiche, and H. S. Olcott, Bull. Environ. Contam. Toxicol., 4, 192 (1969).

63. I. D. Presst, D. J. Jefferies, and N. W. Moore, Environ. Pollution, 1, 3 (1970).

64. S. Ulfstrand, A. Södergren, and J. Rabol, Nature, 231, 467 (1971).

65. G. Westöö and K. Norén, Acta Chem. Scand., 24, 1639 (1970); K. Norén and G. Westöö, Acta Chem. Scand., 22, 2289 (1968).

66. T. Kojima, H. Fukumoto, and S. Makisumi, Jap. J. Legal Med., 23, 415 (1969).

67. V. Zitko, Bull. Environ. Contam. Toxicol., 5, 279 (1970).

68. R. H. deVos and E. W. Peet, Bull. Environ. Contam. Toxicol., 6, 164 (1971).

69. G. E. Bagley and E. Cromartie, J. Chromatogr., 75, 219 (1973).

70. H. O. Sanders and J. H. Chandler, Bull. Environ. Contam. Toxicol., 7, 257 (1972).

71. D. L. Stalling and T. N. Huckins, J. Assoc. Offic. Anal. Chem., 54, 801 (1971).

72. S. Bailey and P. J. Bunyan, Nature, 236, 34 (1972).

73. R. G. Webb and A. C. McCall, J. Assoc. Offic. Anal. Chem., 55, 746 (1972).

74. E. M. Emery and G. M. Gasser, U.S. Pat. 3,520,108 (1970).

75. P. W. Albro and L. Fishbein, J. Chromatogr., 69, 273 (1972).

76. P. Raig and R. Ammon, Arzheim.-Forsch., 20, 1266 (1970).

77. L. Tomori, Magy. Kem. Foly., 76, 437 (1970); CA, 74, 49394 (1971).

78. C. A. Bache and D. J. Lisk, Bull. Environ. Contam. Toxicol., 9, 315 (1973).

79. W. P. Cochrane and A. S. Y. Chau, Advances in Chemistry Series 104, American Chemical Society, Washington, D.C., 1971, Chapter 2.

80. J. W. Dolan, R. C. Hall, and T. M. Todd, J. Assoc. Offic. Anal. Chem., 55, 537 (1972).

81. R. I. Asai, F. A. Gunther, W. E. Westlake, and Y. Iwata, J. Agr. Food Chem., 19, 396 (1971).

82. Application Highlights, No. 24, Polychlorinated Biphenyls, Waters Associates, Framingham, Mass.

83. N. V. Fehringer and J. E. Westfall, J. Chromatogr., 57, 397 (1971).

84. G. K. Westöö, K. Norén, and M. Andersson, Vår Föda, 22, 92 (1970); CA, 73, 86684 (1970).

85. G. Westöö and M. Andersson, Vår Föda, 23, 341 (1971).

86. R. H. deVos and E. W. Peet, Bull. Environ. Contam. Toxicol., 6, 164 (1971).

87. D. L. Stalling and J. N. Huckins, J. Assoc. Offic. Anal. Chem., 56, 367 (1973).

88. M. Beroza, K. R. Hill, and K. H. Norris, Anal. Chem., 40, 1608 (1968).

89. K. D. Bartle, J. Assoc. Offic. Anal. Chem., 55, 1101 (1972).

90. A. C. Tas and R. J. C. Kleipool, Bull. Environ. Contam. Toxicol., 8, 32 (1972).

91. F. J. Biros, A. C. Walker, and A. Medbery, Bull. Environ. Contam. Toxicol., 5, 317 (1970).

92. R. A. Flath, in Guide to Modern Methods of Instrumental Analysis, T. H. Gouw, ed., Wiley-Interscience, New York, 1972, Chapter 9.

93. S. P. Markey, Anal. Chem., 42, 306 (1970).

94. R. Ryhage, Anal. Chem., 35, 759 (1964).

95. F. J. Biros, Residue Rev., 40, 1 (1971).

96. E. J. Bonelli, Am. Lab., February 1971.

97. S. Safe and O. Hutzinger, Chem. Commun., 446 (1971).

98. S. Safe and O. Hutzinger, J. Chem. Soc. Perkin Trans., 1, 686 (1972).

99. S. Safe and O. Hutzinger, J. Chem. Soc. D, 446 (1971).

100. G. E. Bagley, W. L. Reichel, and E. Cromartie, J. Assoc. Offic. Anal. Chem., 53, 251 (1970).

101. E. J. Bonelli, Anal. Chem., 44, 603 (1972).

102. D. G. Shaw, Bull. Environ. Contam. Toxicol., 8, 208 (1972).

103. O. Hutzinger, W. D. Jamieson, and V. Zitko, Nature, 226, 664 (1970).

104. D. L. Stalling, Abstracts of Papers WATR No. 22, 164th National Meeting, American Chemical Society, New York, August 1972.

105. L. O. Ruzo, M. J. Zabik, and R. D. Schuetz, Bull. Environ. Contam. Toxicol., 8, 217 (1972).

106. S. Jensen, A. G. Johnels, M. Olsson, and G. Otterlind, Nature, 224, 247 (1969).

107. K. Veermer and L. M. Reynolds, Can. Field-Naturalist, 84, 117 (1970).

108. V. Zitko, Int. J. Environ. Anal. Chem., 1, 221 (1972).

109. D. L. Grant, W. E. J. Phillips, and D. C. Villeneuve, Bull. Environ. Contam. Toxicol., 6, 102 (1971).

110. P. E. Corneliussen, J. Assoc. Offic. Anal. Chem., 56, 302 (1973).

111. V. Zitko and P. M. K. Choi, Bull. Environ. Contam. Toxicol., 7, 63 (1972).

112. J. W. Rote and P. G. Murphy, Bull. Environ. Contam. Toxicol., 6, 377 (1971).

113. J. A. Armour, J. Assoc. Offic. Anal. Chem., 56, 987 (1973).

114. V. Zitko, O. Hutzinger, W. D. Jamieson, and P. M. K. Choi, Bull. Environ. Contam. Toxicol., 7, 200 (1972).

115. O. Hutzinger, S. Safe, and V. Zitko, Int. J. Environ. Anal. Chem., 2, 95 (1972).

116. O. Hutzinger, W. D. Jamieson, and S. Safe, J. Assoc. Offic. Anal. Chem., 56, 982 (1973).

117. D. L. Stalling, in International IUPAC Congress of Pesticide Chemistry, Tel-Aviv, 1971, A. S. Tahori, ed., Gordon and Breach, New York, 1972, Vol. IV, pp. 413-38.

118. R. G. Webb and A. C. McCall, paper presented at the 164th National Meeting, American Chemical Society, New York, August 1972; J. Chromatogr. Sci., 11, 366 (1973).

119. O. Hutzinger, S. Safe, and V. Zitko, Bull. Environ. Contam. Toxicol., 6, 209 (1971).

120. L. D. Sawyer, J. Assoc. Offic. Anal. Chem., 56, 1015 (1973).

121. S. J. V. Young, C. Finsterwalder, and J. A. Burke, J. Assoc. Offic. Anal. Chem., 56, 957 (1973).

122. G. B. Collins, D. C. Holmes, and F. J. Jackson, J. Chromatogr., 71, 443 (1972).

123. Halowax Chlorinated Naphthalenes, Technical Bulletin, Koppers Co., Inc., Tar Products Division, Pittsburgh, Pa.

124. J. A. Armour and J. A. Burke, J. Assoc. Offic. Anal. Chem., 54, 175 (1971).

125. A. V. Holden, in Marine Pollution and Sea Life, Fishing News (Books) Ltd., Surrey, England, December 1972, p. 2.

126. D. F. Goerlitz and L. M. Law, Bull. Environ. Contam. Toxicol., 7, 243 (1972).

127. D. C. Abbott, J. O'G. Tatton, and N. F. Wood, J. Chromatogr., 42, 83 (1969).

128. D. C. Holmes and M. Wallen, J. Chromatogr., 71, 562 (1972).

129. J. G. Vos, J. G. Koeman, H. L. Vandermaas, M. C. Tennoever deBrauw, and R. H. deVos, Food Cosmet. Toxicol., 8, 625 (1970).

130. V. Zitko, Bull. Environ. Contam. Toxicol., 7, 105 (1972).

131. D. Firestone, J. Ress, N. L. Brown, R. P. Barron, and J. N. Damico, J. Assoc. Offic. Anal. Chem., 55, 85 (1972).

132. Anon., Nature, 226, 309 (1970).

133. G. L. Sparschu, F. L. Dunn, and V. K. Rowe, J. Toxicol. Appl. Pharmacol., 17, 317 (1970).

134. G. R. Higginbotham, A. Huang, D. Firestone, J. Verrett, J. Ress, and A. D. Campbell, Nature, 220, 702 (1968).

135. M. L. Porter and J. A. Burke, J. Assoc. Offic. Anal. Chem., 54, 1426 (1971).

136. J. Ress, G. R. Higginbotham, and D. Firestone, J. Assoc. Offic. Anal. Chem., 53, 628 (1970).

137. Official Methods of Analysis, 11th ed., Association of Official Analytical Chemists, Washington, D.C., 1970, p. 468.

138. D. Firestone, D. F. Flick, J. Ress, and G. R. Higginbotham, J. Assoc. Offic. Anal. Chem., 54, 1293 (1971).

139. D. A. Elvidge, Analyst, 96, 721 (1971).

140. D. T. Williams and B. J. Blanchfield, J. Assoc. Offic. Anal. Chem., 54, 1429 (1971).

141. E. A. Woolson, R. F. Thomas, and P. D. J. Ensor, J. Agr. Food Chem., 20, 351 (1972).

142. T. J. N. Webber and D. G. Box, Analyst, 98, 181 (1973).

143. R. W. Storherr, R. R. Watts, A. M. Gardner, and T. Osgood, J. Assoc. Offic. Anal. Chem., 54, 218 (1971).

144. J. R. Plimmer, J. M. Ruth, and E. A. Woolson, J. Agr. Food Chem., 21, 90 (1973).

145. D. T. Williams and B. J. Blanchfield, J. Assoc. Offic. Anal. Chem., 55, 93, 1358 (1972).

146. A. E. Pohland and G. C. Yang, J. Agr. Food Chem., 20, 1093 (1972).

147. F. L. Mayer, Jr., D. L. Stalling, and J. L. Johnson, Nature, 238, 411 (1972).

148. D. J. Nazir, A. P. Alcaraz, B. A. Bierl, M. Beroza, and P. P. Nair, Biochemistry, 10, 4428 (1971).

149. J. Cerbulis and J. S. Ard, J. Assoc. Offic. Anal. Chem., 50, 646 (1967).

150. G. Ogner and M. Schnitzer, Science, 170, 317 (1970).

151. V. Zitko, Technical Report No. 344, Fisheries Research Board of Canada, St. Andrews, N.B., 1972.

152. V. Zitko, Int. J. Environ. Anal. Chem., 2, 241 (1973).

153. L. Fishbein and P. W. Albro, J. Chromatogr., 70, 365 (1972).

154. O. Schettino and M. I. La Rotonda, Bull. Soc. Ital. Biol. Sper., 45, 1337 (1969).

155. A. A. Chubarova, Vop. Gig. Pitan., 136 (1967); CA, 69, 67953 (1968).

156. L. I. Rapaport, Otkrytiya, Izobret., Prom. Obraztsy, Tovarnye Znaki, 47, 93 (1970); CA, 73, 116184 (1970).

157. M. Wandel and H. Tengler, Deut. Lebensm. Rundsch., 60, 335 (1964).

158. G. Wildbrett, K. W. Evers, and F. Kiermeir, Fette Seifen Anstrichm., 71, 330 (1969).

159. E. Cianetti and M. L. Andriulli, Rass. Chim., 21, 151 (1969).

160. D. L. Stalling, J. W. Hogan, and J. L. Johnson, Environmental Health Perspectives, January 1973, p. 159.

161. R. J. Jaeger and R. Rubin, Science, 170, 460 (1970).

162. W. Pfab, Deut. Lebensmitt. Rdsch., 63, 72 (1967).

163. A. Reichle and H. Tengler, Deut. Lebensmitt. Rdsch., 64, 142 (1968).

164. P. W. Albro, R. Thomas, and L. Fishbein, J. Chromatogr., 76, 321 (1973).

165. W. Bunting and E. A. Walker, Analyst, 92, 575 (1967).

166. P. P. Nair, I. Sarlos, and J. Machiz, Arch. Biochem. Biophys., 114, 488 (1966).

167. G. G. Esposito, Anal. Chem., 35, 1439 (1963).

168. L. Damyanova-Gudeva, Khim. Ind. (Sofia), 43, 393 (1971).

169. M. Swiatecka and H. Zowall, Polimery, 14, 165 (1969).

170. D. A. Nelson, Anal. Biochem., 29, 171 (1969).

171. J. Sliwiok, Microchem. J., 59, 416 (1968).

172. R. E. Majors, J. Chromatogr. Sci., 8, 338 (1970).

173. D. N. Kendall, R. R. Hampton, H. Hausdorf, and F. Pristera, Appl. Spectry., 7, 179 (1955).

Chapter 6

HIGH-PERFORMANCE ELECTROMETER SYSTEMS
FOR GAS CHROMATOGRAPHY

Douglas H. Smith*

Hewlett-Packard Laboratories
Palo Alto, California

---

*Present address: Hewlett-Packard, Avondale, Pennsylvania

I. INTRODUCTION

The flame-ionization detector (FID) has become the most widely used detector in gas chromatography. Together with the electron-capture detector (ECD), these two detector systems account for roughly two-thirds of the new units purchased. Both detectors require special high-impedance amplifiers, or electrometers, and other high-impedance circuits for signal readout. The design of suitable high-performance electrometers requires careful consideration of many parameters not normally encountered at such low levels. In this chapter I discuss these parameters and describe several circuit approaches that have the desired performance. The new electrometers are characterized by an absence of the usual ranging transients and have considerably better range-to-range tracking accuracy than the conventional approaches.

The success of the new transistor-feedback analog electrometers has led to the development of direct digital electrometers. The new digital electrometers are autoranging and directly compatible with digital readout systems.

The design of high-impedance floating bias supplies and pulsers has permitted the use of simpler and more reliable grounded-jet flame detectors and grounded-body electron-capture cells.

Readout systems employing electron-capture detectors also require high-impedance circuit techniques. This chapter describes a readout systems that has improved dynamic range and provides a direct digital output.

II. NOISE LIMITATIONS

Designing circuits to detect the very-low-level signals from the FID and
ECD requires fairly careful assessment of the sources of noise in the
detector, electrometer amplifying semiconductor devices, feedback
components used in ranging, bias supplies, cables, connectors, and
mounting hardware. The last three are difficult to quantify, and acceptable
performance is usually arrived at by evaluation of breadboards and proto-
types. The other sources of noise can be predicted with fair accuracy from
knowledge of resistance or standing current; however, the bandwidth of the
measurement must also be known since the noise in most cases will vary in
proportion to the square root of the bandwidth.

A. Bandwidth Requirements

Reducing the bandwidth does reduce the noise, but a point is reached where
further bandwidth reduction seriously distorts the peak shapes. Figure 1
shows how the peak height, the peak width, and the retention time are
affected by excessive filtering. In this figure peak shapes are shown for
a single-pole (simple resistor-capacitor) analog filter with varying time
constant τ. Most electrometers are designed with a filter time constant
of about 0.1 sec. Peaks with half-height widths $W_{\frac{1}{2}}$ as narrow as 1 sec are
reproduced with good fidelity. This is adequate for all but the very fastest
capillary-column chromatography.

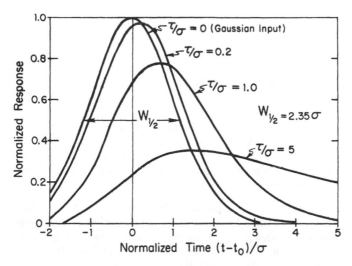

FIG. 1. Effect of filtering on peak fidelity. The peak time is delayed,
the peak height is reduced, but the area remains unchanged.

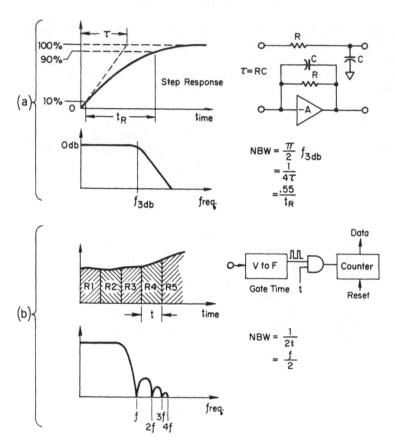

FIG. 2. Noise-bandwidth equations: (a) analog single-pole filter;
(b) digital boxcar integrating filter.

Actually, even though the peak height may be reduced and the peak
width increased with heavy filtering, the area will remain unchanged.   It is
common practice to use more filtering in area-readout devices such as
integrators and computer systems than in instruments in which peak-height
quantitation will be used.

Figure 2 shows how the noise bandwidth (NBW) used in various noise
equations can be determined from measurement or from knowledge of
circuit-component values.   For example, if a simple analog filter with a
time constant of 0.1 sec is to be used, the NBW can be calculated from the
equations shown in Fig. 2a and will be 2.5 Hz.

If a digital filter of the sampled-integrator (or boxcar) type is used with an integrating period t of 100 msec, the corresponding NBW would be 5 Hz (Fig. 2b). A 100-msec period is frequently chosen because it is an exact multiple of both 60- and 50-Hz line frequencies, giving infinite rejection to power-line pickup. The basic sampled-integrator filter is characteristic of the digital electrometers discussed in Section V. If other digital filtering schemes are used, such as curve fitting over a number of points [1], the NBW will usually be the NBW of the basic data points divided by the square root of the number of points used in the fit.

B. Thermal Noise and Shot Noise

In a resistor the kinetic energy of the molecules produces random charge movement, resulting in noise that is called thermal, Johnson, or heat noise. The power available from this random motion is given by

$$P = 4kT \text{ NBW} \quad \text{(watts)} \tag{1}$$

where k is Boltzmann's constant, 1.38×10^{-23} joule/$^\circ$K; T is the temperature in degrees Kelvin; and NBW is noise bandwidth in hertz.

From this equation the root-mean-square (rms) noise developed in a resistor R is given by

$$e_n = (4kTR \text{ NBW})^{\frac{1}{2}} \quad \text{(volts rms)} \tag{2}$$

where R is in ohms. It can also be expressed as a current noise

$$i_n = (4kT \text{ NBW}/R)^{\frac{1}{2}} \quad \text{(amperes rms)} \tag{3}$$

Since the amplitude distribution of the noise is Gaussian, the rms value corresponds to one standard deviation, and it is common practice in gas chromatography to express the noise in terms of its peak-to-peak (pp) value as seen on a recorder

$$i_{n \text{ pp}} = 6i_{n \text{ rms}} \tag{4}$$

Taking the pp value at six times the rms means that the pp noise will fall within its calculated limits 99.7% of the time.

Figure 3 shows a plot of thermal current noise against resistance and can be used to determine performance limits imposed by range and suppression resistors in conventional resistor-ranged electrometers.

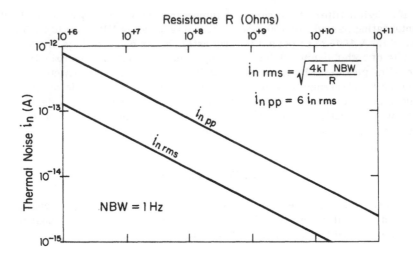

FIG. 3. Thermal current noise for a given resistance.

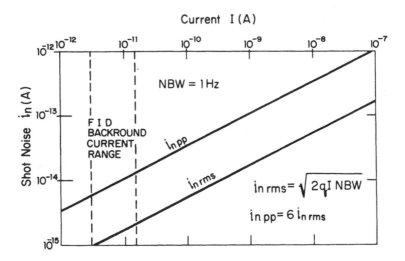

FIG. 4. Theoretical shot noise for a given standing current, showing typical FID background current.

Shot noise is caused by the random emission or collection of current carriers. It is characteristic of leakage currents in semiconductor devices and the standing or background current in the FID and ECD

$$i_n = (2qI\ NBW)^{\frac{1}{2}} \quad \text{(amperes rms)} \tag{5}$$

where q is the charge of the carrier (usually one electron or 1.60×10^{-19} coulomb); I is the standing current in amperes; and NBW is the noise bandwidth in hertz.

Figure 4 shows a plot of shot noise against standing current for an NBW of 1 Hz. Also shown is the typical range of a clean FID where the standing current is from 3 to 15 pA (pA = 10^{-12}A) and the corresponding theoretical shot noise from 6 to 13 fA pp (fA = 10^{-15}A).

In practice contaminated supply gases, high detector temperatures, and column bleed cause an increase in the current noise over the theoretical shot-noise limit by a factor of 2 or more.

C. Field-Effect Transistors

The new electrometers use a field-effect transistor (FET) as a high-impedance amplifier or comparator. The main source of noise in the new electrometers is the shot noise of the FET gate-leakage current (Eq. 5). The gate-leakage current is highly temperature dependent and is also dependent on the operating bias voltages used (Fig. 5). Considerable performance improvement can be achieved by cooling the input FET, and this technique is used in the bipolar analog electrometers described in Section IV.G.

Thermal noise in the channel of the FET appears as a source of voltage noise at the input and will pump against any shunt input capacitance (cable, connector, and FET gate capacitance), giving rise to an apparent current noise [2]

$$i_n(f) = 2\pi f C e_n(f) \tag{6}$$

where f is the frequency in hertz; C is the total input capacitance in farads; and e_n is the voltage noise in volts.

To find the total effective current noise, Eq. (6) should be multiplied by the filter response and integrated over the bandwidth of interest, but for an estimate we can use a 1-Hz bandwidth centered at 1 Hz. For example, suppose we have a system with a 100-pF cable and that no more than a 5-fA

pp current noise can be tolerated.   The maximum allowable voltage noise
due to the input stage would be

$$e_n = i_n/(2\pi fC) = 8 \ \mu V \ pp \tag{7}$$

     This low-voltage noise is easily achieved with a medium-geometry FET;
however, the FET must be operated below ambient temperatures to reduce
the current noise to acceptable levels.   Substantial savings can be realized
by using a small-geometry low-leakage FET and heating the front end.
Section V describes electrometers in which the capacitive current noise has
been reduced by locating the electrometer within a few inches of the detector,
eliminating the cable capacitance and permitting the use of a small-geometry
FET with its somewhat higher voltage noise.   Section VI details the requirements of the temperature-stable zone.

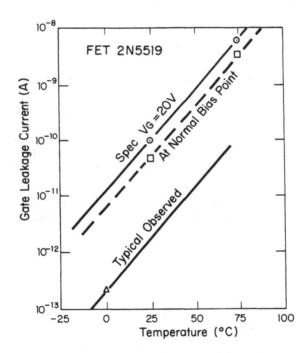

FIG. 5. FET gate-leakage current as a function of temperature.

D. Cable and Connector Noise

In addition to the problems of input capacitance already discussed, cables and connectors are often troublesome sources of noise when flexed or vibrated. When cables are necessary, they should be of a low-noise type and clamped at frequent intervals to the instrument frame.

In new instrument designs the cable and connectors can be eliminated by placing the electrometer front end within a few inches of the detector and making the connection to the collector electrode with a small length of bare wire. This approach is possible with the new electrometers because of their small size and remote-ranging capability.

E. Flame-Ionization-Detector Noise

The FID is basically a shot-noise-limited detector at low sample concentrations; however, at high background levels or with larger samples a gain noise dominates and limits the signal-to-noise ratio (S/N) to about 1000. One way to illustrate this is to plot the S/N as a function of input current (Fig. 6). These curves were obtained by running the detectors on several

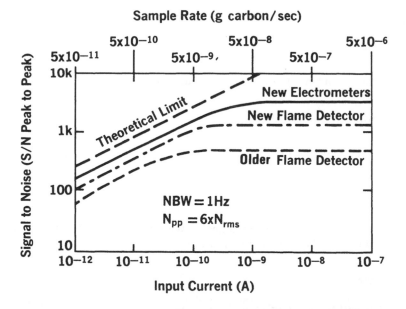

FIG. 6. Signal-to-noise ratios of new electrometers and flame detector as a function of input current (sample rate), compared with those of earlier flame detectors and the theoretical shot-noise limit.

premixed propane-in-nitrogen carrier-gas supplies and taking the ratio of the steady-state response to the pp noise. The gain noise appears to come from short-term temperature fluctuations in the flame. Note that the new electrometers are always better than the detectors and that the electrometers are close to the theoretical shot-noise limit at the lower currents. The gain noise in the electrometers is due to short-term temperature changes of approximately $0.01°C$ in the front-end thermal zone.

A chromatogram made with one of the new analog electrometers and using a new FID is shown in Fig. 7. The detector base-line noise, not to be confused with several small impurity peaks, is below 15 fA pp. Since the detector sensitivity is 20 mC/g carbon, the system can detect samples as small as 1.5 pg carbon/sec (S/N = 2).

F. Electron-Capture-Detector Noise

In the ECD, ^{63}Ni is used as a source of high-energy β particles. Each β particle generates about 100 secondary electrons, which quickly reach thermal energies. The random decay of the radioactive ^{63}Ni source is the primary source of noise. Since the curie is defined as 3.7×10^{10} disintegrations per second, a typical ^{63}Ni source of 15 mCi has a disintegration rate K of 5.6×10^8 β particles per second.

The total number of β particles in a time period t is given by

$$N_\beta = Kt \tag{8}$$

Since they are random, the variance or rms noise is

$$n_\beta = (N)^{\frac{1}{2}} \tag{9}$$

The S/N from above will be

$$(S/N)_{rms} = (Kt)^{\frac{1}{2}} \tag{10}$$

Substituting for the 15-mCi source and taking a time of 0.5 sec (NBW of 1Hz), we find that the expected rms S/N is

$$(S/N)_{rms} = 1.7 \times 10^4 \tag{11}$$

The pp S/N will be one-sixth the above value

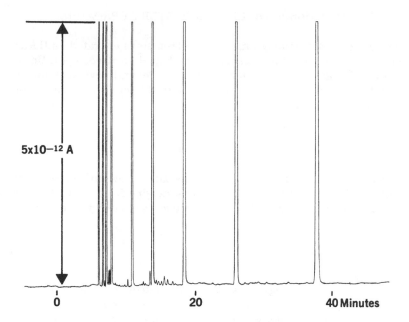

5x10⁻¹² A

0   20   40 Minutes

FIG. 7. Typical chromatogram generated using transistor-feedback electrometer and new FID (sample was n-paraffin blend on a 150-ft capillary column with instrument set to range 1, Atten. ×1, -hp- 5711A gas chromatograph).

$$(S/N)_{pp} = 2.8 \times 10^3 \tag{12}$$

The signal in the above expressions is the base-line value of the frequency or current in the two ECD systems described in Section VIII. For example, in Fig. 8 the base-line frequency of the new ECD system was 453 Hz. Since the theoretical S/N is 2800, the expected base-line noise would be 0.16 Hz, which is close to the measured value.

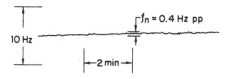

FIG. 8. Electron-capture-detector base-line noise (-hp- 5709, NBW 1Hz, detector temperature 300°C, flow 60 ml/min, argon-10% methane).

III. COMPARISON OF ELECTROMETER APPROACHES

In prior gas-chromatography practice electrometers operated as linear
current-to-voltage converters using a very-high-impedance operational
amplifier with high-value feedback resistors. Several new electrometers
based on transistor feedback have been designed to circumvent many of the
problems of the conventional approach.

A. Resistor Feedback

The conventional approach uses a high-impedance, low-noise amplifier in
conjunction with several feedback resistors ranging from 10^6 to 10^{11} ohms
(Fig. 9a). The magnitude of the output voltage is related to the input
current by

$$V_{out} = R_f I_{in} \qquad\qquad\qquad (13)$$

To be compatible with integrators, the flame detector is biased in such a
way that the output of the electrometer is positive with increasing sample.
An additional resistor and a variable voltage supply are connected to the
input to suppress the flame background and electrometer offset currents.
Also present, but not shown, is a balance adjustment used to compensate
for the voltage offset of the electrometer amplifier so that the zero on the
recorder remains unaffected by changes in the recorder attenuator or
range-selector switch.

The high-value resistors tend to have poor stability with time and
temperature as well as poor tolerance, limiting the repeatability of
measurements and causing large range-to-range tracking errors. Furthermore, range changing is done at high impedance levels, causing large
transients (Fig. 10a). The switching mechanism also increases the
number of possible leakage paths and takes up space.

The traditional technique puts severe requirements on the noise performance of the operational amplifier. Adding to Eq. (13) the terms for the
feedback-resistor thermal current noise i_{nR}, the amplifier current noise,
i_{nA}, and the amplifier voltage noise e_{nA}, the output voltage is given by

$$V_{out} = R_f I_{in} + \left[(i_{nR} R_f)^2 + (i_{nA} R_f)^2 + (e_{nA})^2 \right]^{\frac{1}{2}} \qquad (14)$$

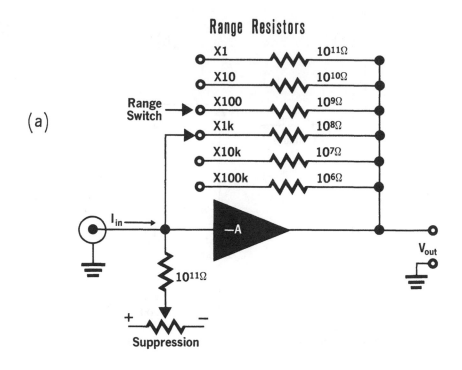

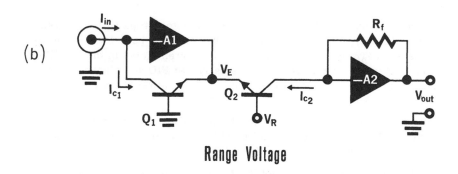

FIG. 9. Electrometer approaches: (a) conventional resistor feed-back; (b) new transistor feedback.

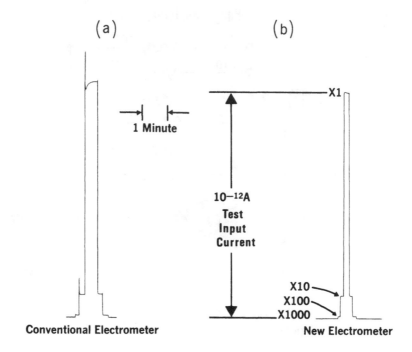

FIG. 10. Range-switching transients: (a) resistor-feedback electrometer has large transients; (b) new electrometer approach does not have
measurable transients.

Since the current noise in the feedback resistor drops with increasing
resistance (Eq. 3), a suitably high-value resistor can be selected to make
the noise term i_{nR} small compared to i_{nA}. Current noise in the amplifier
then limits the most sensitive range. Ideally, an FID and electrometer
readout system should span a total dynamic range of nearly nine orders of
magnitude (10 fA pp noise to 5 μA full scale).

To achieve the 10-fA pp noise level, all leakage currents must be held
below 10 pA (Eq. 5) and the parallel combination of the feedback and suppression resistors must be well above 10^{10} ohms (Eq. 3).

Voltage noise in the amplifier limits the dynamic range on the less
sensitive ranges (R_f small). The dynamic range on range $\times 10$ and higher
ranges must be greater than 10^6 to be directly compatible with integrators
and computer systems. It is difficult and expensive to design an amplifier
with both low current noise and low voltage noise primarily because small-
geometry semiconductor devices that have low current noise have high
voltage noise, and large-geometry devices with low voltage noise have high
current noise.

B. Transistor Feedback

To circumvent many of the problems in conventional designs, a new approach was taken (Fig. 9b). It uses a logarithmic amplifier to span the wide dynamic range of the input current. The linearity of the overall response is then restored by an exponential converter following the logarithmic input stage [3].

The circuit is based on the exponential relationship between the magnitude of the forward-bias emitter voltage V_E and the magnitude of the collector current I_c of transistors

$$I_c = I_s \left[\exp \left(\frac{qV_E}{nkT} \right) - 1 \right] \tag{15}$$

where I_s is the reverse saturation current in amperes; q is the charge on the electron, 1.60×10^{-19} coulomb; V_E is the emitter voltage (base grounded, sign plus for forward bias and negative for reverse bias) in volts; n is a constant near unity; k is Boltzmann's constant, 1.38×10^{-23} joule/$^\circ$K; and T is the temperature in degrees Kelvin.

For forward-biased junctions the –1 term can be neglected. Assuming the input amplifier of Fig. 9b is ideal, V_E of the log transistor Q1 is related to the input current I_{in} by

$$V_E = \left(\frac{nkT}{q} \right) \ln \left(\frac{I_{in}}{I_{S1}} \right) \tag{16}$$

The emitter-base voltage of the exponential transistor Q2 is increased by the ranging voltage V_R, and assuming the output amplifier of Fig. 9b is ideal

$$V_{out} = I_{S2} R_f \exp \left[\frac{q(V_E + V_R)}{nkT} \right] \tag{17}$$

Substitution for V_E yields the overall response

$$V_{out} = I_{in} R_f \left(\frac{I_{S2}}{I_{S1}} \right) \exp \left(\frac{qV_R}{nkT} \right) \tag{18}$$

Note that the response is linear with respect to I_{in}. The ratio I_{S2}/I_{S1} is constant, as is q/nkT, because Q1 and Q2 are maintained at a constant temperature.

Ranging is effected by adjusting V_R in discrete steps (59 mV/decade or 18 mV/octave range). Since the ranging is done at a low-impedance point away from the input node, ranging transients are completely eliminated (Fig. 10b). The circuit is easily calibrated by the addition of a small offset to V_R.

Because of the impedance transformation between stages, the operational amplifiers need not have both low current noise and low voltage noise. The input stage must have low current noise, but because the input is operating in essentially a current mode, its voltage noise does not appear in the output. The current noise is held low by the use of a very-small-geometry FET.

The output stage must have low voltage noise, but its current noise is insignificant because of the low impedance of this stage. The voltage noise is held low by the use of a large-geometry FET.

Unlike conventional resistor-feedback electrometers, which have range-to-range tracking errors of 2 to 10%, the new transistor-feedback electrometers track within a fraction of 1%. Response factors (i.e., calibration with a known sample) derived on one range can be used with confidence on other ranges.

Since the ranging is a set of low-impedance voltage steps, remote-ranging capability is provided in the new analog electrometers and auto-ranging is designed into the digital electrometers.

IV. ANALOG TRANSISTOR-FEEDBACK ELECTROMETERS

A few additional topics concerning transistor-feedback electrometers will be covered before describing actual circuit approaches. These include noise from input devices, suppression techniques, linearity limitations of the log-exp approach, response speed limitations, and range-voltage stability requirements.

A. Input-Stage Noise

The voltage noise of the input-stage amplifier (Fig. 9b) can be ignored if the collector-base junction impedance R_{CB1} of the log feedback transistor Q1 is sufficiently high. For the small-geometry FETs used, the voltage noise measured is about 12 μV pp at a 1-Hz NBW. This voltage noise e_{nFET} will cause an input current noise

$$i_n = \frac{e_{nFET}}{R_{CB1}} \tag{19}$$

For this current noise to be less than 5 fA pp, R_{CB1} must be greater than

$$R_{CB1} = \frac{e_{nFET}}{i_n} = 2.4 \times 10^9 \text{ ohms} \tag{20}$$

The zero-bias R_{CB} of transistors can be measured directly or can be calculated by measuring the collector-base leakage current at 1 V of bias. Note that all of the devices shown in Fig. 11 have an R_{CB} well in excess of the required value at room temperature, and thus this potential source of current noise can be ignored. If the devices are cooled to near $0^\circ C$, R_{CB} will increase by a factor of 10, and the medium-geometry transistors 2N3964 and 2N3904 can be used. However, if the front end is heated to near $50^\circ C$, R_{CB} will decrease by a factor of 10, and only a transistor of the very smallest geometry, such as the 2N4044, is acceptable.

Two sources of shot noise are found in the front-end stage. The gate leakage of the input FET gives rise to a shot noise and was discussed in Section II.C. In the cooled versions larger geometry FETs can be used (Fig. 5); however, in the heated versions only the very-small-geometry devices, such as the 2N5909, will have sufficiently low current noise.

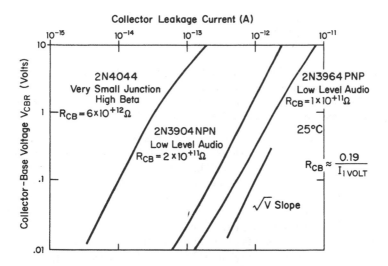

FIG. 11. Collector leakage current for several transistors and empirical formula for calculation of the zero bias collector-base impedance.

The other source of noise is the shot noise in the collector current of the log feedback transistor. Since the collector current equals the input current, this source of current noise degrades the theoretical S/N at low currents by a factor of $\sqrt{2}$. Note that in Fig. 6 the new electrometers are roughly $\sqrt{2}$ below the theoretical shot noise limit for the FID ion current but still above the best flame detector that has been constructed and evaluated.

The output-stage voltage noise must be held low for a wide dynamic range compatible with integrators. The 2N5519 FET used has sufficiently low voltage noise for a 10-V full-scale, 10^6 dynamic range output stage.

B. Suppression Techniques

All of the analog electrometers require a method of suppressing the flame background and input offset currents. In conventional electrometers this is done with an additional high-value resistor at the input and adjustable voltage source (Fig. 9a).

In the new transistor-feedback electrometers the suppression current is supplied to the input by a transistor connected as an adjustable current source (Fig. 12a). The suppression current I_{C3} is an exponential function of the forward bias suppression voltage V_S

$$I_{C3} = I_{S3} \left[\exp\left(\frac{V_S q}{nkT} \right) - 1 \right] \tag{21}$$

Because this arrangement is shot-noise limited in the same manner as the source, it can cover a much wider current range than was available through a single high-value resistor that has a fixed current noise and maximum value limited by the adjustable voltage source (Fig. 9a).

In the heated versions, in which the decreased input shunt impedance of several collector-base junctions cannot be tolerated, the suppression can be moved to the output stage (Fig. 12b). However, now the suppression current must be ranged by the same multiplier that is used in the exponential output stage. For example, suppose the suppression was 10% on range ×1; then it must become 1% on range ×10 to prevent the base line (zero) from moving when ranging. The suppression and ranging can be made to track in the proper fashion by connecting the emitter of Q3 to V_R since the collector current in Q2 and Q3 are both ranged by V_R

$$I_{C2} = I_{S2} \exp\left(\frac{V_E q}{nkT} \right) \exp\left(\frac{V_R q}{nkT} \right) \tag{22}$$

and

$$I_{C3} = I_{S3} \exp\left(\frac{V_S q}{nkT}\right) \exp\left(\frac{V_R q}{nkT}\right) \tag{23}$$

C. Linearity Limitations

Up to now we have assumed that the factor n was a constant near unity in the basic equation for the collector current

$$I_C = I_S \left[\exp\left(\frac{qV_E}{nkT}\right) - 1 \right] \tag{15}$$

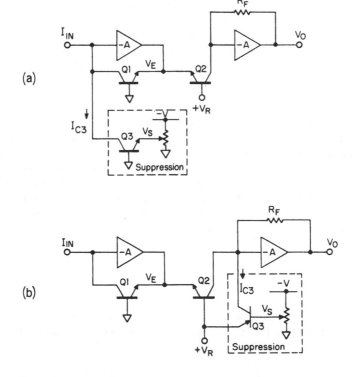

FIG. 12. Suppression circuits: (a) input stage; (b) output stage, which requires range tracking.

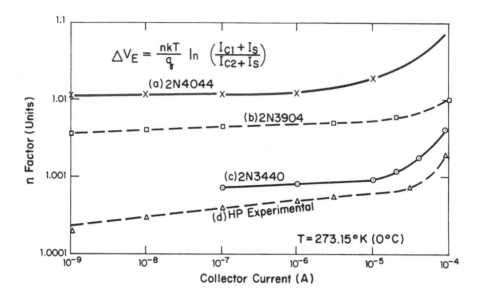

FIG. 13. Log conformity factor n versus collector current for several transistors: (a) 2N4044 small geometry, very low leakage; (b) 2N3904 medium geometry, low-level audio; (c) 2N3440 high voltage, medium geometry; (d) HP experimental, very wide base.

However, the factor n is a minor function of collector current resulting from a change in the effective base width of the transistor as the emitter-base junction is forward biased (Fig. 13). This effect is small for current levels up to 10 μA but does result in a slight nonlinearity of approximately 0.2% of reading per decade of input current (Fig. 14a). Since the change in n is nearly linear with V_E (log I_C), its effect can be reduced to 0.05% per decade by feeding back a small portion (1/500) of V_E to the base of Q1 (Fig. 14b). This improved linearity is nearly an order of magnitude better than the FID.

D. Response Speed

The time constant of the conventional electrometer is related to the feedback resistor R_f the effective shunt capacitance C_f, and is independent of the input bias current (Fig. 2a)

$$\tau = R_f C_f \tag{24}$$

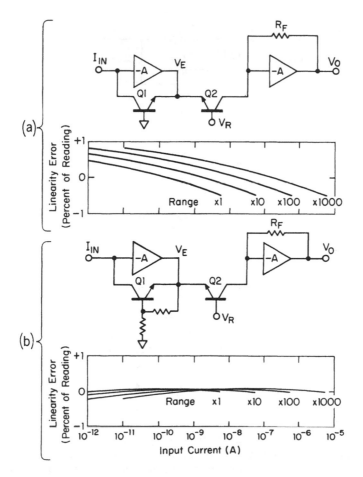

FIG. 14. Nonlinearity and range-to-range tracking accuracy: (a) basic circuit; (b) improved circuit with feedback to base of log transistor.

Unlike the conventional electrometer, the new transistor-feedback electrometer has a response speed that is a function of the input current (Fig. 15a). From the basic equation for the log feedback transistor

$$I_C = I_S \left[\exp \left(\frac{qV_E}{nkT} \right) - 1 \right] \tag{15}$$

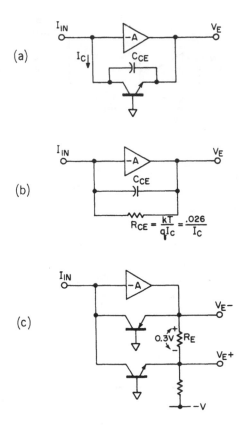

FIG. 15. Speed limitation of the log-exp feedback approach: (a) stray collector-emitter capacitance limits speed; (b) equivalent small-signal model; (c) circuit for fast bipolar electrometers by overlapping plus and minus log stages.

The change in collector current ΔI_C for a small change in emitter voltage ΔV_E, is given by

$$\Delta I_C = I_S \left(\frac{q}{nkT} \right) \exp\left(\frac{qV_E}{nkT} \right) \Delta V_E \qquad (25)$$

The exponential term can be eliminated by substituting the steady-state value I_C and neglecting the -1 term from Eq. 15

$$\frac{\Delta I_C}{\Delta V_E} = \frac{qI_C}{nkT} \qquad (26)$$

The effective collector-to-emitter resistance R_{CE} at room temperature $(T = 300^\circ K)$ is

$$R_{CE} = \frac{\Delta V_E}{\Delta I_C} = \frac{nkT}{qI_C} = \frac{0.026}{I_C} \qquad (27)$$

and can be used in a small-signal model (Fig. 15b). The collector-to-emitter capacitance C_{CE} times R_{CE} from Eq. 27 yields the effective time constant

$$\tau = \frac{0.026C_{CE}}{I_C} \qquad (28)$$

For single-polarity electrometers the input current will always be at least as great as the flame background current (typically no lower than 5 pA), and C_{CE} is usually no more than 0.5 pF; thus from Eq. 28 the maximum time constant would be

$$\tau = 2.6 \times 10^{-3} \quad \text{sec} \qquad (29)$$

and will decrease with increasing input current. This is about an order of magnitude faster than the convention electrometers with comparable noise performance. For a bipolar electrometer the input current will be near zero when suppressed and from Eq. 28 the time constant would increase to a very high value. To maintain fast response near zero input current, a constant voltage drop across R_E overlaps the plus and minus feedback transistors, preventing the collector current from dropping below 10^{-13} A (Fig. 15c).

E. Range-Voltage Stability

The range voltage V_R must be held relatively constant for good gain stability, and the range steps must also be fairly accurate for accurate range-to-range tracking. The basic equation for the transistor-feedback electrometer derived in Section III.B is

$$V_{out} = I_{in} R_f \left(\frac{I_{S2}}{I_{S1}} \right) \exp \left(\frac{qV_R}{nkT} \right) \qquad (18)$$

From Eq. 18 the sensitivity to small changes in V_R is given by

$$\frac{\Delta V_{out}}{V_{out}} = \Delta V_R \left(\frac{q}{nkT}\right) \tag{30}$$

and near room temperature

$$\frac{\Delta V_{out}}{V_{out}} = \frac{\Delta V_R}{0.0026} \tag{31}$$

Thus for a S/N of 1000, V_R must have less than 26 μV pp noise, and for 1% range-to-range tracking, the range steps in V_R must be accurate to 260 μV. Since the total span in the range voltage for four decades of range is 240 mV, 1% range-to-range tracking implies 0.1% accuracy in the range digital-to-analog converter (DAC). This is no problem since we are dealing with relatively low impedance levels in the DAC and also the DAC components can be temperature stabilized along with the other front-end components.

F. Single-Polarity Analog Electrometer

The FID background current is always in the same direction as the signal current, and when the collector bias is negative with respect to the grounded jet, the detector current is also in the same direction as the gate-leakage current of the input FET. Thus the input current will never go through zero, and it becomes possible to build a transistor electrometer with only a single-polarity log and exp stage. Figure 16 shows a simplified schematic of a single-polarity electrometer designed for use in a low-cost single-detector gas chromatograph. The suppression is done at the exponential output stage to maintain the highest possible input impedance. Only one transistor, Q1, is connected to the input node, permitting operation at a fixed high temperature of 50°C. Four decade ranges are provided by a switch that increases V_R in four uniform steps of approximately 62 mV.

Since the output stage is also temperature regulated, the balance control need be adjusted only occasionally to compensate for long-term drift. Not shown in Fig. 16 is the filter to limit the NBW and the recorder attenuator. Actually the recorder attenuation could be combined with the ranging by providing additional 18-mV steps (×2 attenuation), but then the electrometer would not be directly compatible with integrators.

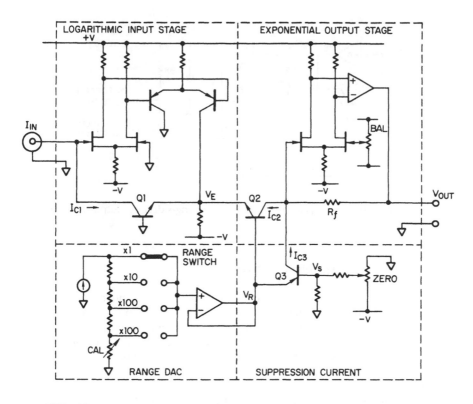

FIG. 16. Simplified schematic of a single-polarity transistor-feedback analog electrometer.

Another approach to the single-polarity electrometer eliminates the output amplifier and adds a floating power supply to allow for a 10-V swing on the collector of the exponentiating transistor Q2 (Fig. 17). The dynamic range on a given range is limited by the thermal noise of the 10-kilohm output load resistance at the low end and by 10 V at full scale. The thermal noise for a 10-kilohm resistor is 80 nV pp at a 1-Hz NBW (Eqs. 2 and 4). Thus the potential dynamic range of this approach would be greater than 10^8. The disadvantage of this approach is that there is some interaction between the recorder, integrator, and computer outputs because of the high output impedance, whereas with the exponential stage amplifier the output imped-ance is near zero (Fig. 16).

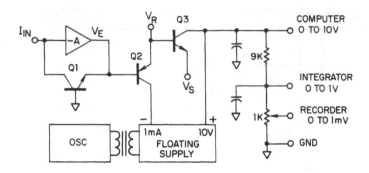

FIG. 17. Single-polarity high-dynamic-range direct-output analog electrometer.

G. Bipolar Analog Electrometer

When dual flame detectors are used in the compensating mode, the output current may be of either polarity and near zero at base line. Figure 18 shows a simplified schematic of a bipolar electrometer designed to operate over a range of from below 10-fA pp noise to ±5 μA at full scale [4]. Matched pairs of NPN and PNP log-exp transistors permit bipolar operation. The suppression current can also be of either polarity. Range-control input can be either contact closures or standard 5-V logic levels, permitting remote computer control of the range without transients (Fig. 10b).

The expanded scale in Fig. 19a shows the noise to be about 3.5 fA pp; however, the electrometer noise is not visible on the recorder trace during normal operation of the instrument on the most sensitive range (Fig. 19b).

The stability of the new electrometer and FID is exceptional, drifting less than 100 fA over a 24-h period (Fig. 20). This particular electrometer was designed for good performance with a fairly large input capacitance from a cable. As a result, a medium-geometry FET with relatively low voltage noise is used for the input. To cope with the high gate leakage of the FET and the low input shunt impedance of the four transistors, the front end is cooled to near 0°C with a thermoelectric cooler, thus reducing the gate-leakage current and increasing the shunt impedance by a factor of 10. Figure 21 shows some of the mechanical detail of the mounting and cooling of the front end. Note that the front end is in a hermetically sealed container, and all feed-through headers are mounted on the hot side of the cooler, completely eliminating potential humidity problems. The thermal design is very important and is discussed in Section VI. Figure 22 shows the relative mounting of the major components in the electrometer plug-in frame. Because of the exceptional stability, the familiar balance control has been eliminated from the front panel.

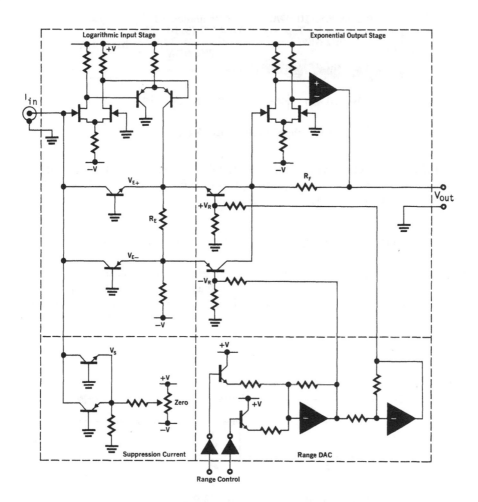

FIG.18. Simplified schematic of a bipolar analog electrometer.

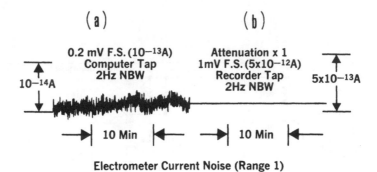

(a) (b)

0.2 mV F.S. (10^{-13}A)   Attenuation x 1
Computer Tap   1mV F.S. ($5x10^{-12}$A)
2Hz NBW   Recorder Tap
10^{-14}A   2Hz NBW $5x10^{-13}$A

10 Min   10 Min

Electrometer Current Noise (Range 1)

FIG. 19. Bipolar analog electrometer current noise: (a) expanded
scale shows noise to be about 3.5 fA pp; (b) trace during normal operation
on range 1, attenuation ×1.

V. DIGITAL TRANSISTOR-FEEDBACK ELECTROMETERS

With the rapidly decreasing cost of digital logic circuits and the availability
of inexpensive small processors, it is now feasible to design gas chromatographs that are completely digital. The peak integrator, which has in the
past been a separate instrument, can now be integrated into the gas chromatograph.

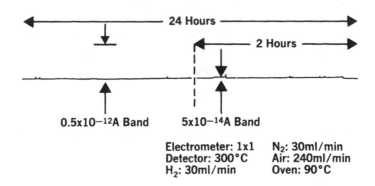

24 Hours

2 Hours

0.5x10^{-12}A Band   5x10^{-14}A Band

Electrometer: 1x1 N_2: 30ml/min
Detector: 300°C   Air: 240ml/min
H_2: 30ml/min   Oven: 90°C

FIG. 20. Stability of the bipolar analog electrometer with the FID on.

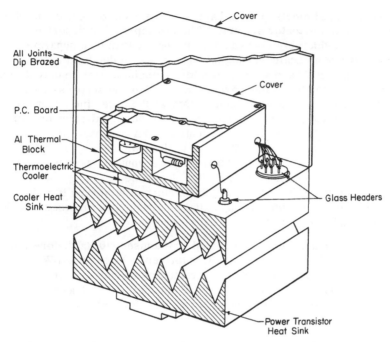

FIG. 21. Mechanical detail of cooled front end used in the bipolar analog electrometer.

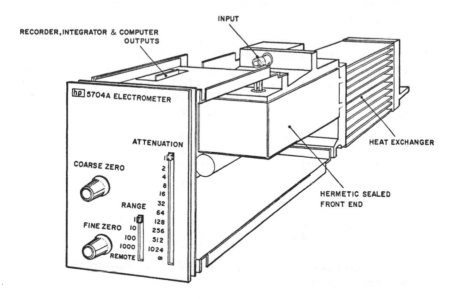

FIG. 22. Bipolar analog electrometer plug–in module for FID gas chromatograph.

Several digital electrometers have been designed to meet the need for compatible digital detector systems. The two approaches described in this section cost considerably less and perform better than do a separate analog electrometer and an analog-to-digital converter. In the digital electrometers the input FET and a single feedback transistor are the only devices that are critical for low-noise performance. Also, no suppression circuit is required since the processor can remember the base-line offset and subtract it from subsequent data points. Since the digital electrometers can be ranged automatically without transients and during a run without errors, the full dynamic range of the FID can be utilized without any operator decisions as to the optimum range. All front-panel controls have been removed, which permits the mounting of the digital electrometer very close and directly behind the FID. For dual-column compensated runs two single-polarity detectors and autoranging digital electrometers are used, the data points being subtracted in the instrument processor.

Readings are averaged for a period of 100 msec (10 Hz) before transfer to the processor in which a digital filter further reduces the NBW to near 1 Hz. Because both of the digital electrometer approaches have relatively low input capacitance (no cables), a small-geometry (low-leakage) FET can be used, and the front end can be heated to 50°C for stable operation.

A. Current-to-Frequency Digital Electrometer

Since both the FID and the collector of the feedback transistor Q1 look like current sources, the input shunt capacitance C of the electrometer can be used as a passive integrator (Fig. 23a).

The FID is biased in such a way that the current flows into the electrometer charging C in the positive direction (Fig. 23b). When the capacitor voltage V_C passes through zero, the zero comparator output goes positive, switching a feedback pulse to the emitter of Q1 at the next clock, F_o. The emitter voltage pulse V_E causes a current pulse I_C in the collector, discharging the capacitor back to the negative side of zero.

Since charge is conserved at the input (ideal integrator) and since the loop maintains the input voltage near zero, the time-average collector current must equal the time-average input current over any extended time period T

$$\frac{1}{T} \int_0^T I_C(t)dt = \frac{1}{T} \int_0^T I_{IN}(t)dt \tag{32}$$

The term $I_C(t)$ is either zero or at a fixed value controlled by V_E when the analog switch SW is on.

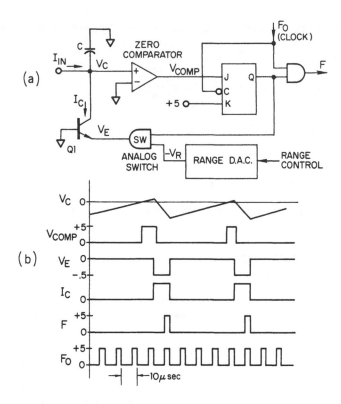

FIG. 23. Current-to-frequency digital electrometer front end: (a) block diagram; (b) timing diagram.

Recalling the basic equation for I_C

$$I_C = I_S \left[\exp\left(\frac{qV_E}{nkT}\right) - 1 \right] \tag{15}$$

When SW is off, $V_E = 0$ and $I_C = 0$. When SW is on, $V_E = V_R$ and the collector current is given by

$$I_C = I_S \left[\exp\left(\frac{qV_R}{nkT}\right) - 1 \right] \tag{33}$$

The time-average value of the collector current $\hat{I}_C$ is simply its maximum value (Eq. 33) times its duty cycle (ratio of the on time to the total time). For the current-to-frequency converter the duty cycle is simply F/F_0 and is limited to 50% of maximum. Neglecting the -1 term, we have

$$\hat{I}_C = \frac{1}{T} \int_0^T I_C(t)dt = I_S\left(\frac{F}{F_0}\right) \exp\left(\frac{qV_R}{nkT}\right) \qquad (34)$$

Since this must equal the time-average input current

$$\hat{I}_{in} = \hat{I}_C = I_S\left(\frac{F}{F_0}\right) \exp\left(\frac{qV_R}{nkT}\right) \qquad (35)$$

Solving for the output frequency F, we obtain

$$F = \hat{I}_{in} \frac{F_0}{[I_S \exp(qV_R/nkT)]} \qquad (36)$$

Note that F is linearly related to the input current and scaled by an exponential function of V_R. The terms I_S and q/nkT are constants because the feedback transistor and other front-end components are maintained at a constant temperature near $50^\circ C$. Adjusting V_R in discrete steps of about 18 mV with the 4-bit range DAC provides the 16 binary ranges necessary to cover the 10^9 dynamic range of the detector. In one version that was investigated the autoranging was done at a 1-kHz rate and permitted tracking of signals moving as fast as one octave every 1 msec without loss of area information. Note that this electrometer is continuously integrating the input, and when peak areas are determined by summing data points, the calculated area is equal to the true integral and not an approximation as occurs in conventional sampled data systems.

In Eq. 36 the factor n is independent of the duty cycle and dependent only on V_R. The nonlinearity pattern encountered with the analog electrometers (Fig. 14) is not present in the new digital electrometers. The slight change in n with V_R is fully compensated by increasing the 18-mV range steps roughly 0.2% when calibrating.

The main disadvantage of this converter is the limited maximum clock frequency of about 200 kHz because the rise and fall times of the emitter voltage pulse affect the fixed-area collector current pulse. The current-to-time converter described next largely circumvents this problem and provides a factor-of-10 improvement in resolution on a given range.

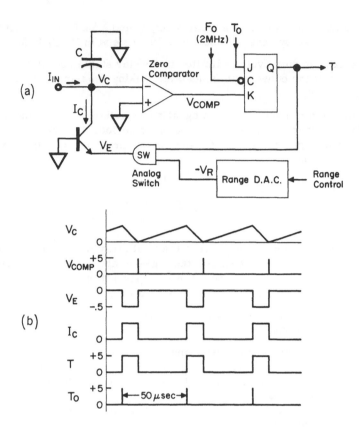

FIG. 24. Current-to-time digital electrometer front end: (a) block diagram; (b) timing diagram.

B. Current-to-Time Digital Electrometer

The operation of the current-to-time digital electrometer (Fig. 24) is similar to that of the current-to-frequency converter.

The feedback current is turned on at a fixed periodic time interval T_0. When I_C has discharged the voltage on C to zero, I_C is turned off on the following clock pulse F_0. The output time interval T can be shown to be linearly related to the input current

$$T = \hat{I}_{in} \frac{T_0}{[I_S \exp (qV_R/nkT)]} \tag{37}$$

As with the frequency-to-current converter, I_S and q/nkT are constants. Ranging is provided by several discrete voltage steps in V_R as before.

The maximum duty cycle is limited to 50% for stable operation. At levels above this the conversion is unstable, taking two periods to produce the correct output.

The required minimum input integrating capacitance to prevent saturation of the input stage is related to the basic period, $T_0 = 50\ \mu\text{sec}$, the maximum input current, $I_{in} = 5\ \mu\text{A}$, and the maximum input voltage of the comparator, $V_C = 10\ \text{V}$, by

$$C = I_{in} \frac{T_0}{2V_C} = 12.5\ \text{pF}. \tag{38}$$

This turns out to be just about equal to the total stray input capacitance. Since the input capacitance is small, the high-voltage noise of a small-geometry low-leakage FET can be tolerated, permitting the use of a heated rather than cooled, temperature-stable zone.

The comparator must be designed to handle a 10-V input yet retain fast response near zero input.

The range DAC and analog switching system must have a sufficiently low impedance for accurate operation at the maximum input current of $5\ \mu\text{A}$. Since the duty cycle is limited to 50%, the peak emitter current will be $10\ \mu\text{A}$ ($I_E \approx I_C$ for high-β transistors) and since a drop in the emitter voltage, ΔV_E, of $260\ \mu\text{V}$ will cause a 1% error, the maximum allowable range DAC impedance is

$$R_{DAC} = \frac{\Delta V_E}{I_E} = 26\ \text{ohms}. \tag{39}$$

This impedance level is easy to achieve either by designing a low-impedance DAC or by using an operational amplifier. If an amplifier is used, the voltage noise must be held below $26\ \mu\text{V}$ pp for a S/N of 1000.

Since the duty cycle never exceeds 50%, there will be a 25-μsec period between each reading in which ranging can be done. Up-ranging occurs whenever the duty cycle is greater than $3/8$, and down-ranging occurs at $3/16$ duty cycle. Since the noise will be relatively high at the 20-kHz digitizing rate (10-kHz NBW) and the ranging occurs in binary steps without hysteresis, the range-to-range tracking errors tend to be smoothed out when readings are averaged for the 1-Hz NBW. Smoothing of the range-to-range tracking errors helps considerably in the slope detection of small peaks.

VI. THERMAL DESIGN

For stable operation of both analog and digital transistor electrometers, careful consideration must be given to thermal design.

A. Input Field-Effect Transistor

The temperature dependence of the input FET gate-leakage current I_G can result in excessive base-line wander. From Fig. 5 the temperature dependence of I_G is roughly 7% per degree Celsius and can be expressed as

$$\frac{\Delta I_G}{I_G} = \frac{\Delta T}{13} \tag{40}$$

Solving for the required temperature stability, we obtain

$$\Delta T = 13 \left(\frac{\Delta I_G}{I_G} \right) \tag{41}$$

From shot-noise considerations I_G must be below 10 pA for a 10-fA pp noise level (Fig. 4). From Eq. 41 a 5-fA pp wander contribution with $I_G = 10$ pA implies a temperature stability of

$$\Delta T = 6.5 \times 10^{-3} \ {}^{\circ}C \tag{42}$$

The required stability is achievable by mounting the FET near the thermistor temperature sensor in a temperature-controlled zone using an aluminum or copper block.

Since the front end is essentially current mode, the FET offset-voltage temperature coefficient will not appear in the output.

B. Log-Exp Transistors

The collector current I_C is a function of temperature for a constant emitter bias voltage V_E. Stating it in reverse, at constant I_C, V_E will vary roughly 2 to 3 mV/${}^{\circ}$C, depending on the current density and particular process used in making the transistor. Figure 25 shows how the emitter temperature coefficient, tempco, varies with the emitter-base voltage. In the analog electrometers there is some reduction in the effective tempco because of the temperature tracking between the log and exp transistors. The worst case occurs on range ×1, where typically the ratio of current densities

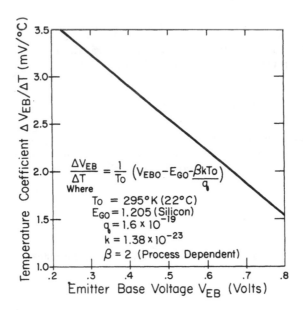

FIG. 25. Temperature dependence of emitter-base voltage for constant collector current (V_{EBO} measured at T_o).

between the exp and log stage is 10^4. Since the tracking is 100 $\mu V/^\circ C$ per decade, the worst-case tracking will be 400 $\mu V/^\circ C$. From the basic equations it can be shown that the temperature tracking affects the gain stability in the same way as does the range-voltage stability

$$\frac{\Delta V_{out}}{V_{out}} = \frac{\Delta V_R}{0.026} \tag{31}$$

For an S/N ratio of 1000 the tracking must be within 26 μV, and since the worst-case tracking is 400 $\mu V/^\circ C$, the temperature regulation must be better than

$$\Delta T = \frac{26 \ \mu V}{(400 \ \mu V/^\circ C)} = 0.065^\circ C \ pp \tag{43}$$

The log and exp transistors must be located in close proximity to each other to prevent any temperature differences between them. Since the tempco of the log stage will be near 3 mv/$^\circ$C (Fig. 25) a temperature difference of only 0.01°C will give rise to a 0.1% change in gain.

An additional potential problem is the power dissipation in the exp transistor near full scale. If this power dissipation causes a temperature rise of 0.01°C, a 0.1% error will occur in the output. The temperature rise ΔT is related to the junction-to-case thermal impedance θ_j and the power dissipation P by

$$\Delta T = \theta_j P \tag{44}$$

The exp-transistor collector current at full scale is typically 50 μA and the total voltage drop is near 0.7 V, the product being the power dissipation (P = 35 μW). For $\Delta T = 0.01$°C the required thermal impedance is

$$\theta_j = \frac{\Delta T}{P} = \frac{286°C}{watt} \tag{45}$$

Since the θ_j of typical transistors ranges from a high of 300°C/watt for TO-18 case-style low-level transistors to a low of 1°C/watt for TO-3 power transistors, proper device selection can circumvent this power-dissipation problem.

The power dissipation in the log transistor of the analog electrometer and the feedback transistor of the digital electrometers is lower by a factor of 10 than in the above example and can be ignored.

C. Temperature-Controlled Zones

From the preceding discussion in this section it is obvious that the front-end components must be maintained at a fixed stable temperature with a short-term stability (thermal noise) of slightly better than 0.01°C.

Two basic thermal designs have been used with the new electrometers. For the low-cost single-polarity analog electrometers and the autoranging single-polarity digital electrometers a heated zone near 50°C has been successfully used (Fig. 26a). For the high-performance bipolar analog electrometer a thermoelectrically cooled thermal zone is preferred (Fig. 26b).

The main advantages of the heated zone over the cooled version are lower cost, lower power consumption, and smaller mechanical size; furthermore, the hermetic sealing of front-end components is not necessary for high-humidity operation. The disadvantage is that the selection of front-end components is somewhat limited because of increased leakage currents at the elevated temperature. During the last few years significant advances have been made in low-leakage FETs and transistors permitting the use of heated zones.

To achieve the low thermal noise of 0.01°C a thermistor sensor is used in a low-noise bridge with a proportional plus integral control error amplifier. All critical front-end components are mounted in a small die-cast aluminum block and insulated from fast ambient-temperature changes.

Both designs take about 10 min from initial turn-on to reach stable operation. This time is more than adequate, because in normal gas-chromatograph operation the detector heated zones take much longer to stabilize.

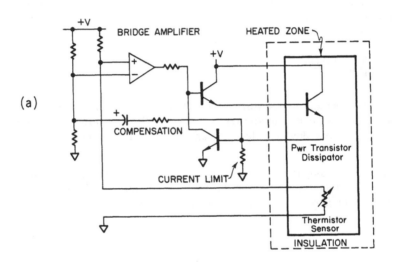

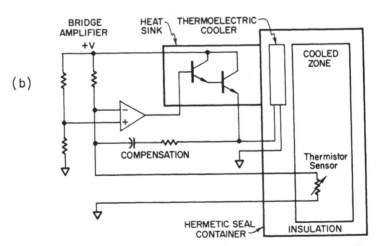

FIG. 26. Simplified schematic of two temperature controllers: (a) heated zone using power transistor; (b) cooled zone using thermoelectric cooler.

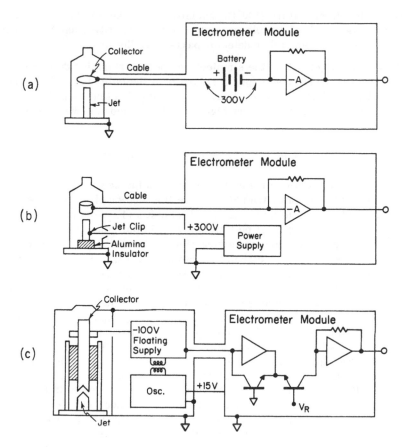

FIG. 27. Flame-detector biasing arrangements: (a) earliest designs, which used a grounded jet and battery in series with the electrometer; (b) isolated jet later adopted to eliminate troublesome battery; (c) most recent design, which reverts to grounded jet and replaces battery with a floating bias supply.

VII. FLOATING BIAS SUPPLIES

The very first flame detectors had a grounded jet. A floating battery in series with the electrometer was used to generate a field between the jet and collector electrode for ion collection (Fig. 27a). This approach had its problems primarily with cable noise and the limited shelf life of the battery. The isolated jet FID replaced the battery with a ground-referenced 300-V supply and improved performance significantly (Fig. 27b). However, it requires a high-temperature (400°C) insulator on the jet and is less desirable for a wide range of sample linearity due to competitive ion collection between the collector and other detector parts near the jet.

The grounded-jet FID is the simplest and best performing detector but does require a floating bias supply in series with the collector and electrometer input (Fig. 27c). Two new floating bias supplies that meet the requirements of low noise and very high isolation impedance were designed.

A. Single Transformer, Bipolar Supply

Because of the large collector electrode design in the new FID, the polarizing voltage need only be greater than 80 V. The floating supply is required to have a very high isolation impedance ($>10^{11}$ ohms) and a very low voltage noise. The voltage noise must be kept below 0.5 ppm since it can pump against the collector capacitance, giving rise to current noise. These requirements are met by the circuit diagrammed in Fig. 28a. A 2-Mhz oscillator coupled by a small toroid transformer to a floating tuned secondary with two voltage doublers provides both a positive and a negative bias voltage. With an analog bipolar electrometer, this allows two detectors to be connected in a differential mode for gas-chromatograph runs that use a compensating column. Each floating-supply output passes through a two-pole filter that reduces the voltage noise to levels below 0.5 ppm in a 1-Hz bandwidth. The capacitors must be of the plastic film type. Other types of capacitors, particularly electrolytics, show frequent abrupt changes in the output voltage, which with the collector capacitance results in large current spikes. To achieve the high isolation impedance the primary is insulated with a high-quality dielectric from the floating and guarded secondary circuit (Fig. 28b). The cable capacitance is used as a simple RC filter to reduce the small amount of high-frequency coupling from the primary turns to an acceptable level of a few millivolts.

For single-detector instruments where a single-polarity analog electrometer is used the +100-V secondary circuit and polarity switches are eliminated.

B. Dual Transformer, Single-Polarity Supply

Another approach that uses two toroid cores and a one-turn link (Fig. 29a) was developed for use primarily with the digital electrometers. Lower capacitive coupling between the oscillator and secondary was necessary for proper operation of the input comparator in the digital electrometers.

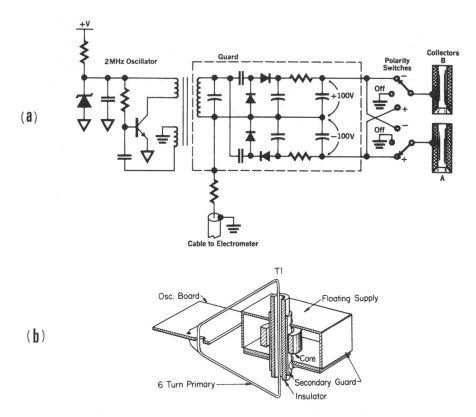

FIG. 28. Bipolar floating bias supply: (a) simplified schematic; (b) construction detail of floating and guarded secondary windings and transformer.

This approach is somewhat easier to assemble but does use an additional core (Fig. 29b). When used with digital electrometers, the oscillator is locked to the same high-frequency clock that is used for timing to prevent any cross modulation and its resulting low-frequency noise. The secondary winding is tuned by adjusting C2 for maximum current in the output-regulating zener diode by monitoring the voltage drop across the 100-ohm current-sampling resistor. All components in the secondary circuit should be guarded to reduce microphonics, to eliminate spiking caused by the electrostatic collection of small dust particles, and to eliminate voltage drops across possible leakage paths to ground.

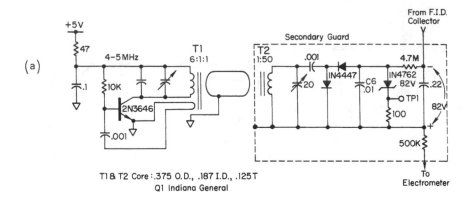

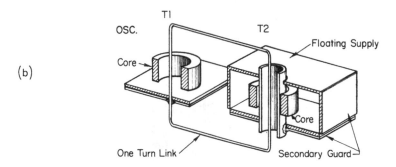

FIG. 29. Dual transformer floating bias supply: (a) simplified schematic; (b) construction detail of two transformer cores — guarding and one-turn coupling link.

VIII. ELECTRON-CAPTURE-DETECTOR SYSTEMS

Up until recently the ECD was pulsed at a fixed periodic rate, and the time-average current output was monitored with the same elctrometer designed for FID systems (Fig. 30a). The relatively high background current (1 to 10 nA) was suppressed at the input by substitution of a lower value suppression resistor. For the best sensitivity long pulse periods of 100 μsec to 1 msec were used, but for linearity at high sample concentrations short pulse periods of 10 to 100 μsec were necessary. At a given fixed pulse rate the linear dynamic range was usually limited to less than 1000.

Another approach that has extended the dynamic range to 10^5 uses a variable pulse rate (Fig. 30b) [5]. The pulse rate is varied by a feedback arrangement to maintain the time-average current out of the cell constant

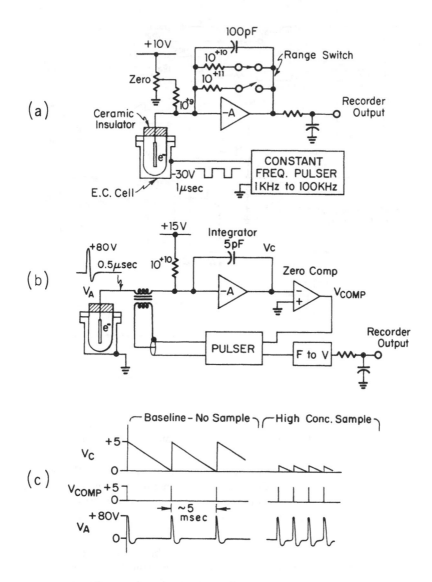

FIG. 30. Comparison of ECD readout systems: (a) conventional approach, which used FID electrometer and constant pulse rate; (b) new constant–current linear readout system, which was variable–frequency pulser; (c) timing diagram for new system.

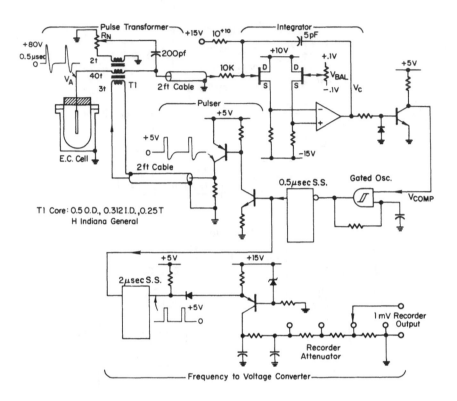

FIG. 31. Simplified schematic of constant-current linear ECD system.

at about 1.5 nA for a 15-mCi ^{63}Ni source. It can be shown that the output frequency is linearly related to the sample concentration. Another change that was made placed the pulse on the anode electrode with a transformer permitting the design of a grounded body cell. This simplified the construction of the cell and improved reliability.

Near base line (no sample) the frequency will be low since the charge collected at each pulse interval is high. When a sample is present, some of the free electrons in the cell are captured and the charge per pulse decreases. The feedback will increase the frequency to maintain the time-average current out of the cell constant at 1.5 nA.

From Section II. F we see that the theoretical maximum S/N of a 15-mCi ECD is 2800, and for a standing base-line current of 1.5 nA it follows that the current noise will be roughly 0.5 pA pp at a 1-Hz NBW. Note that this is greater by a factor of 50 than the 10-fA pp level in the FID. As a result, just about any FET can be used for the ECD front-end integrator and adequate stability can be achieved without temperature regulation. Figure 31 shows a simplified schematic of the new ECD system.

For proper operation the ECD cell anode capacitance must be neutralized to prevent overloading of the integrator. Neutralization is achieved by capacitively coupling the input node to an additional secondary winding that is out of phase with the anode pulse. Adjustment with R_N is made for the proper waveform at the integrator output (Fig. 30c). The pulse-transformer secondary winding is insulated from the core with a low-leakage dielectric, and the assembly is mounted at the detector.

For digital readout systems the pulser frequency is counted for a fixed period of 100 msec and then passed to the processor. In analog instruments the frequency is converted to a voltage, filtered to a 1-Hz NBW, and then outputed to a recorder attenuator. Since the total dynamic range is 10^5, ranging is not necessary in ECD systems and only the conventional recorder attenuator is provided.

A gated oscillator is used in the feedback system to prevent "lock-up" during power turn-on, and the balance control is used to null out the contact potential of the ECD cell.

ACKNOWLEDGMENTS

The author is indebted to several individuals from Hewlett-Packard's Avondale Division at Avondale, Pennsylvania, who contributed material to this chapter and were responsible for the excellent design of the production units.

Ernest Zerenner, John McFadden, and Hans Van Heyst were responsible for the bipolar analog electrometers. The bipolar floating bias supply and the single-polarity analog electrometer were developed by William Buffington. The high-performance flame detectors were done by Roger Nalepa. Ernest Zerenner and Jeff Fromm developed the new digital electrometers. James Sullivan was responsible for the new electron-capture-detector cell and the constant-current linear readout system.

REFERENCES

1. A. Savitzky and M. Golay, Anal. Chem., 36, 1627 (1964).

2. E. J. Kennedy, A Study of the Theoretical and Practical Limitations of Low-Current Amplification by Transistorized Current-Feedback DC Electrometers, ORNL TM 1726, Clearing House, Springfield, Va., April 1967, pp. 43-44.

3. D. H. Smith, Hewlett-Packard J., March 1973, p. 2.

4. D. H. Smith, R. B. Bump, and E. H. Zerenner, paper presented at the 1972 Pittsburgh Conference, Cleveland, Ohio.

5. J. J. Sullivan and D. H. Smith, paper presented at the 1972 Pittsburgh Conference, Cleveland, Ohio.

Chapter 7

STEAM CARRIER GAS-SOLID CHROMATOGRAPHY

Akira Nonaka

Institute for Optical Research
Kyoiku University
Tokyo, Japan

I. INTRODUCTION

As is well known, gas chromatography (GC) can be classified into two
categories: gas-solid chromatography (GSC) and gas-liquid chromatography
(GLC). The former has hardly been used unless the samples to be analyzed
are gaseous at room temperature. However, since GSC has many advantages [1, 2], many attempts have been made to overcome its shortcomings,

principally by improving the stationary solids used in the column so that high-boiling and/or polar samples can be analyzed [3-6].

For the mobile phase, only inert gases, such as helium and nitrogen, have been used ever since the GC method was established. Active gases, which may interact with sample molecules in the gas phase, have not been used as the mobile phase, probably because it was not thought that the interaction between sample and carrier-gas molecules could be strong enough to force elution of the sample molecules from the stationary phase.

If the mobile phase in GSC is an inactive gas and the stationary phase is an ordinary inorganic porous-powder adsorbent, then the samples, which are generally liquid and solid at room temperature, are not normally eluted at column temperatures below about 300°C. This may be because the adsorption energies of liquid and solid samples on such adsorbents are so high (tens of kilocalories per mole) [7, 8] that the sample molecules are too firmly adsorbed to be eluted. It was found, however, that even with the inorganic porous adsorbents used in gas-solid chromatography, water vapor (steam) as the carrier gas can easily elute samples that are liquid or solid at room temperature, even high-boiling and/or very polar organic materials [9-20].

Water vapor has been tried by many chromatographers as a component of a mixed carrier gas to deactivate the surface of stationary solids [21] or supports for stationary liquids [22, 23], or with water as the stationary liquid [24, 25]. Few have used water vapor or steam alone as the carrier gas. In GSC especially it has been a commonplace that the last trace of water must be completely removed or reproducibility of sample elution will suffer.

It has been well recognized that steam as carrier gas promotes rapid elution of high-boiling samples. However, it has not yet been directly established that any interactions between water and sample molecules in the gas phase force elution of the sample molecules. On the other hand, it is known that a vapor (i.e., a gas below the critical temperature) such as steam is adsorbed and forms a multilayer at pressures not much lower than the saturation pressure at that temperature [26]. The presence of a multilayer of adsorbed water under conditions similar to those employed in steam carrier gas-solid chromatography (SSC) has been proved experimentally (A. Nonaka, unpublished data). However, little experimental evidence of multilayer adsorption of steam at high temperatures (100 to 150°C) and high pressures (1 to 4 atm) has been reported.

It is reasonable to assume that in most cases the effect of steam carrier gas in promoting the rapid elution of various types of sample is due to the multilayer adsorption of water, which decreases the adsorption energies of sample molecules to the stationary surface, rather than to any molecular interactions between water and sample molecules in the gas phase.

Especially in the case of high-boiling hydrocarbon samples, which are caused to elute more rapidly by steam carrier gas, it is not always reasonable to expect strong intermolecular interactions between steam and samples in the gas phase.

There are many active gases that can be used in GSC in addition to steam. Typical ones are the vapors of formic acid and hydrazine hydrate, which have been employed very effectively in GSC of organic acid and amine samples, respectively, although both are diluted with water vapor to reduce their corrosiveness [19, 20].

Since an adsorbed multilayer is formed on the surface of the stationary solid when the vapor of a liquid is used in as a mobile GSC phase, this type of GC may be assumed to be on the borderline between GSC with a stationary solid deactivated with the vapor and GLC with a low-loaded liquid column. It is not yet definitely determined whether the sample molecules are adsorbed on the stationary phase or whether they are dissolved in the surface layer.

The fact that carrier and sample molecules in the gas phase may interact by physical or chemical means such as dipolar interaction or hydrogen bonding cannot be disregarded. This is especially true in the case of formic acid or hydrazine hydrate vapor carrier gas and extremely polar sample materials. In this matter, SSC may resemble liquid chromatography, in which the interaction between mobile phase and sample molecules is rather significant and the kind of mobile phase chosen is important.

Vapors as carrier gases also have a significant effect on sample elution in GLC [27], but reproducible chromatographic operations may be rather difficult because any active carrier gas can strip the stationary phase.

II. OUTLINE OF STEAM CARRIER GAS-SOLID CHROMATOGRAPHY

Steam carrier gas-solid chromatography has various significant features and advantages that depend mostly on its being a sort of GSC using an active carrier gas as mobile phase. They are as follows:

1. The tailing effect is insignificant even with extremely polar samples

2. High-boiling sample materials can be eluted at a comparatively low column temperature

3. Retention times are mostly rather short, as though they were on a lightly loaded column in GLC; this means that SSC may be suitable for trace analyses, since the shorter the retention time, the larger the peak height with a flame-ionization detector (FID)

4. Operation of the GC system at a high column temperature is rather easy, because inorganic heat-resistant phases are used in the column

5. Aqueous samples, either solutions, emulsions, or suspensions, can be analyzed by direct injection without any pretreatment, such as extraction, concentration, or derivatization, since the GC system is always filled with water

6. The hydrogen-flame-ionization detector, a highly sensitive detector, can be used to advantage in combination with steam carrier gas

7. The columns used in SSC are not subject to deterioration from water.

It is a distinct feature, among the various advantages mentioned here, that dilute aqueous solutions of very polar organic samples can be rather easily analyzed using SSC with FID.

Analysis of aqueous samples by either ordinary GSC or GLC has in general been difficult. In the case of GSC, even a minute amount of residual water in the carrier gas causes a considerable change in the properties of the stationary phase surface, due to strong adsorption on the surface, and results in poor reproducibility of the elution peaks. On the other hand, in most cases in GLC, the water in the sample is eluted and displays a significant tailing, which may either mask the sample peaks or significantly reduce the reproducibility of peak separation. Furthermore, water in the samples accelerates the deterioration of most organic liquids used as stationary phases. This effect increases as the amount of water increases and is particularly troublesome for trace analyses of dilute aqueous samples. There is another reason for the practical impossibility of direct GLC of extremely dilute aqueous solutions (10 ppm or less) of organic substances: the stationary liquid is eluted to some extent along with the water of the sample either as vapor or as hydrolyzed products, and the resulting peaks mask the sample peaks. The situation cannot be improved by the use of an inorganic stationary phase since no suitable inorganic material has yet been found. These difficulties, however, can be easily solved by adopting SSC, a form of GSC in which steam carrier gas is used with an ordinary adsorbent such as silica, alumina, or magnesia.

Many substances improve SSC results when added to the steam. For example, the addition of formic acid or hydrazine hydrate (or ammonia) gives an excellent carrier gas in the SSC of organic acid or amine samples, respectively. In these cases FID can still be employed. If a column temperature below $100^{\circ}C$ is needed for SSC of fairly low-boiling samples, steam as a carrier gas can be introduced after being diluted with an inert gas such as helium and nitrogen. Otherwise the steam could condense on the adsorbent surface at the column temperature. In this case water vapor containing an inert gas retains its ability to elute polar samples from the inorganic porous adsorbent.

In addition to the various advantages already mentioned, it has been found that most vaporizable organic samples, at least such as can be analyzed by ordinary GC, can also be analyzed by SSC, although in some cases the SSC method seems to be somewhat inferior to ordinary GC in terms of the number of theoretical plots (NTP) and resolution. This is probably because SSC is new and has not been studied as much as the traditional GC. It may be expected that in the near future, when SSC has been more fully studied, it will become a general-purpose, rather than merely a specialized, method.

III. INSTRUMENTAL CONSIDERATIONS

The apparatus for SSC resembles that for traditional GC in many aspects except that the former has a steam generator for the carrier steam. Commercially available chromatographs may be modified so that the carrier steam will not condense anywhere. The sample-injection port, the detector, and the connecting portion between the detector and the column end should be of glass because SSC is most often used for extremely polar or high-boiling samples.

A. Steam Generator

There are two useful methods for obtaining a steady flow of steam carrier gas. One is to use a steam boiler capable of generating superheated steam with a control valve to regulate the flow. A second method is that of "column head vaporization" of water to provide the carrier steam. In the latter case water or a water solution of carrier materials other than water is pumped at a constant flow rate into the column head, or a vaporizing port, maintained at a temperature above the boiling point of the carrier. The former method has been applied to SSC from its beginning and can give chromatograms with comparatively low levels of detector noise. The second method is the better in several respects. By utilizing a liquid pump and a small evaporator, the steam-generating apparatus becomes rather simple and can easily be applied to introduce a mixed-vapor carrier gas by pumping the corresponding carrier liquid mixture. Furthermore, carrier-vapor flow rates can be controlled precisely by using a constant-flow-rate liquid pump.

1. Steam Boiler Method

The capacity of the boiler used as a steam generator should be 100 to 200 ml since the flow rate of the steam carrier gas is less than about 100 mg/min and the columns used are less than 3 mm in internal diameter. The stainless-steel boiler is guaranteed for a pressure of 10 kg/cm² and is thermostated by an electric heater so as to generate steam at a constant pressure of about 2 to 3.5 atm. The heater of the boiler is controlled by detecting either the pressure or the temperature change of the boiler so that a constant pressure of steam is obtained. Simple SSC can be carried out by merely supplying a constant heat flow to the boiler (by a constant voltage or current) without any precise control system. The regulating valve used to control the steam flow rates and the tube connecting the valve and the column head must be heated evenly to a temperature above that of the boiler. When a minute amount of organic material is to be analyzed in the SSC system, the steam generated in the boiler passes through a heated column (500 x 5 mm i.d.) packed with cupric oxide pellets (analytical reagent) in order to remove organic impurities in the steam, although in ordinary use the cupric oxide column can be eliminated by using once-distilled water as the boiler water. Flame-ionization detectors can function at signal levels below 1×10^{11} A full scale, that is, at a noise level below about 5×10^{-14} A, if the steam generator, the steam tubes, and the other parts of the chromatograph are made thermally very stable.

Since ordinary flow meters are unsuitable for measuring the steam flow, other devices are employed. The author has used a dew-point hygrometer placed within the exhaust chimney of the FID. The meter was calibrated against the amount of water vapor produced by a pure hydrogen flame whose flow rate was precisely known.

2. Liquid Pump Method

Liquid pumps for carrier materials must operate at a rather high pressure (up to about 5 atm) to obtain a steady flow of carrier vapor against the column pressure drop. A constant-flow-rate liquid pump for SSC can be constructed from a syringe (10 or 20 ml) and a screw rod geared to push the plunger at 0.01 to 0.3 mm/min, giving a flow rate of 2 to 100 μl/min. The syringe must be of quartz or borosilicate glass since the water for the carrier steam may be mixed with a very corrosive solute, such as formic acid or hydrazine hydrate. For the same reason, Teflon (Du Pont) tubing of less than 0.5 mm i.d. is used to carry the liquid from the pump to the vaporizing port. Fairly smooth evaporation of the carrier steam may be expected if the pipe and vaporization-port inlet are extremely narrow or are tightly packed with silica wool.

In place of a mechanical pump a system consisting of a water container, needle valve, and high-pressure gas cylinder can be used (see Fig. 1c). Such a system can easily be improvised by the chromatographer to supply a vapor carrier gas to a conventional GC system for laboratory use. Since the gas cylinder is necessary only to provide the pressure to drive the liquid against the analytical-column pressure drop, almost any inert gas at a

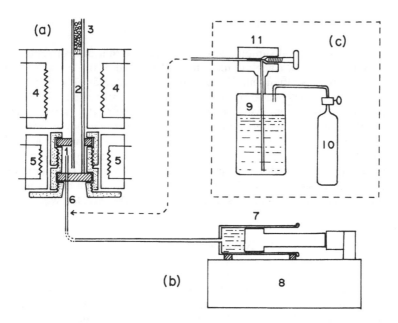

FIG. 1. System for supplying carrier vapor: (a) column head and carrier-vapor-generating portion; (b) constant-flow-rate liquid pump; (c) another type of pumping system. Key: 1, carrier-vapor-generating port; 2, sample-injection port; 3, analytical column; 4, electric heater for vaporizing carrier and sample; 5, electric heater for rubber septum; 6, capillary tube for leading carrier material; 7, glass syringe; 8, driving mechanism; 9, water container; 10, high-pressure gas cylinder; 11, needle valve.

suitable pressure can be used unless it is quite soluble in the liquid. This is a simple method for obtaining steam carrier gas, but it is rather difficult to measure the flow rate precisely in this method.

In order to obtain a continuous flow of carrier vapor from the water or water solution pumped into the SSC system, a vaporizing port must be employed. A typical vaporizing port might have a volume of 5 to 10 cm^3 and be packed with silica granules (about 30 mesh) and maintained at 150 to 180°C by an electric heater. The head of the analytical column or the sample-injected port, maintained usually at a rather high temperature, also serves as a vaporization port if the liquid is pumped directly into the head of the analytical column, as seen in Fig. 1.

It is not difficult to supply the heat necessary to vaporize the liquid and maintain the column head at a constant temperature to ensure a steady flow rate of carrer vapor, since the flow of liquid is usually very small and the heat required for vaporization is also small, at most a few watts.

B. Injection Port

As in ordinary GC, liquid samples can be injected with a microsyringe through a silicone-rubber septum used to seal the end of the injection port. The silicone-rubber septum and the sample-evaporation area should be heated by separate, independently placed, electric heaters to avoid damage to the silicone rubber from overheating and the action of the carrier vapor, and to avoid condensation of carrier vapor. When water vapor mixed with formic acid or hydrazine hydrate vapor is introduced as a carrier gas, corrosion of the injection port may be rather severe. In most cases borosilicate glass can be used for the inside of the sample-vaporizing port and high-quality stainless steel for the other parts of the column head, unless this would expose a large area of stainless steel to the carrier vapor.

C. Analytical Column

Various kinds of inorganic porous powders are used as stationary solid phases (adsorbents) in SSC. Suitable adsorbents are prepared from inorganic substances, such as alumina, silica-alumina, silica, and magnesia, which are chosen for their thermal stability even in the presence of water vapor and other carrier substances. Few organic adsorbents have been used in SSC, because organic materials, even those of high molecular weight, are eluted comparatively easily with the carrier gas whose principal component is water, formic acid, or hydrazine hydrate. Carbon and graphite can also be used in some cases, although they have peculiar water-adsorption properties. The specific area of the adsorbents may range from several square meters to several hundreds of square meters per gram, according to the sample to be analyzed. The adsorbents should be sieved as evenly as possible in a certain mesh size between about 50 and 120 mesh. In some cases commercially available adsorbents can be used as the solid phase without any modifications. In other cases they must be modified by some inorganic salts, acids, or bases, sometimes after being calcined at a high temperature, to make them suitable for the effluent conditions of the samples to be analyzed. The nature of the adsorbent surface can often be greatly changed by such pretreatment. Some conventional adsorbents prepared for GSC and supports for GLC can be used in SSC either "as is" or after slight modification. Generally speaking, an activated adsorbent having a rather high specific area is necessary for separation of nonpolar samples or samples of low molecular weight, whereas lower surface activity and smaller specific area are needed for highly polar samples and those having high boiling points. As a rule, for samples that may be easily hydrolyzed by the water vapor, adsorbents with strongly acid sites must not be used. Furthermore, the acidity or the basicity (in an extended sense) of the adsorbents must be taken into

account to avoid chemisorption of sample molecules. Porous glass and
silica having no silica-gel-like surface are typical adsorbents with weak
surface reactivity.

Almost all kinds of vaporizable organic materials can be analyzed by
SSC. Studies on the effectiveness of stationary solids in SSC analyses
have not been as extensive as studies on stationary liquids in GLC. Some
of the separations obtained in SSC can be inferior to those obtained in GLC.
It is expected, however, that this inferiority will certainly be overcome by
additional studies on SSC in the near future as SSC becomes more common.
This expectation is also based on the fact that some solid adsorbents have
been mixed with stationary liquids in GLC in order to improve separation
[28, 29]. This fact means that some solid-phase adsorbents, which differ
in their ability to adsorb sample components, give better separation than
do liquid stationary phases that differ in their ability to dissolve the sample
components.

In some cases SSC elution peaks are somewhat wider than those obtained
with a well-designed ordinary GLC. This may be ascribed to the fact that
the extremely fine and deep pores in the adsorbent powder particles require
a long diffusion time for sample molecules; as a result the eluted peak is
wider, and larger HETP values are found in the chromatograms [30, 31].
In GLC these fine and deep pores are partly filled by the stationary liquids.
This explanation is borne out by the fact that sharper elution peaks are
obtained when Porasil F (Waters Associates, Inc.), whose pores are
comparatively evenly sized and are not as fine as those of other ordinary
adsorbents, is employed as a stationary solid of SSC. The most desirable
shape for adsorbent particles to be used in SSC is spherical with a non-
porous core and thin, uniform porous surface layer, such as the supports
developed for high-speed liquid chromatography [32]. Rapid mass transfer
is possible since diffusion of sample molecules occurs only in the thin layer
on the surface. Development of this kind of adsorbent seems to be very
important for the advance of SSC.

Table I shows typical adsorbents (stationary solids) that have been used
in SSC with fairly satisfactory results. Pretreatments of the adsorbents
used in SSC are as follows: Adsorbent powders are washed, as a general
practice, with mineral acid solutions (20% hydrochloric acid and 35% nitric
acid) before use. Some diatomaceous adsorbents are somewhat improved
in their separating ability by a hydrofluoric acid modification that is
carried out by keeping the powders immersed in 10% hydrofluoric acid for
about an hour. The immersion of diatomaceous firebrick in a 20% aqueous
solution of a fluoride (KF or KHF_2) is so effective that polyalcohol samples
can be rapidly eluted from the adsorbent. A phosphoric acid modification,
for columns used in analyzing acidic samples, is made by baking the
adsorbent powder at 500°C with a small amount (3% of the powder by weight)
of phosphoric acid for about 5 min so that the acidity of the column will last

for a prolonged period. In this case the phosphoric acid treatment must also be applied to the borosilicate glass column.

Borosilicate glass tubing is usually suitable for the column, but stainless steel or aluminum can be used in some cases, unless they are corroded by the carrier or sample materials, or unless the solute molecules are adsorbed strongly on the columns.

The usual analytical columns used in SSC are 2 to 4 m long. When a stationary solid of comparatively low surface area, such as Corning GLC-100, is used, the column length must be 4 m or more.

The diameter of the column used in SSC has to be determined by taking into consideration the sample size to be injected, although, as in ordinary GLC, the smaller the column diameter, the higher the column efficiency. The minimum permissible diameters have been obtained experimentally on some analytical columns used in SSC. For example, the permissible values of the diameter are about 1.8, 2.5, and 3.7 mm for 0.6-, 1.5-, and 5.0-μl samples, respectively in the case of diatomaceous firebrick column in analyzing alcohols. These values seem to be rather large in comparison with the case of GLC; that is, the maximum sample size permissible in SSC is not as large as in GLC on the same diameter of column. When samples larger than those mentioned above are used, the resolution of the eluted peaks is decreased or the number of theoretical plates is reduced. This is probably because a solid stationary phase cannot hold as much sample material (the adsorbate amount that can obey Henry's law) as a liquid stationary phase having a fairly thick liquid layer. This phenomenon becomes significant in the case of adsorbents with a low specific surface area, such as glass beads.

D. Air Bath

The temperature control of the air bath used in SSC is not as difficult as in the case of ordinary GLC. The SSC chromatograms show no sign of base-line fluctuations or base-line drifts that could be ascribed to the vaporization of the stationary phase and to an imprecise temperature control of the column since no organic materials, whose elution is sensed by the FID, are used in SSC.

Temperature programming in SSC is as effective as in ordinary GC for a sample whose components' boiling points range widely. In this case the base-line drift is also insignificant since there is no elution or organic materials that would increase with column temperature.

TABLE I

Solid Phases (Adsorbents) Used in SSC

Sample	Adsorbent
Low-boiling hydrocarbons	1. Activated alumina 2. Molecular sieves
High-boiling hydrocarbons	1. Diatomaceous firebrick (Isolite LBK-28,[a] C-22, Chromosorb P AW,[b] etc.) 2. Porous silica (e.g., Porasil F[c])
Low-molecular-weight monohydric alcohols, aldehydes, ketones, ethers, and esters	1. Diatomaceous firebrick (Isolite LBK-28, C-22, Chromosorb P AW, etc., all modified with hydrogen fluoride) 2. Porous silica (e.g., Porasil F) 3. Silica-alumina deactivated by adding a small amount of inorganic salts
Organic acids and phenols	1. Diatomaceous firebrick (Isolite LBK-28, Chromosorb P AW, Shimalite,[d] etc., all baked with 3% phosphoric acid) 2. Deactivated diatomaceous firebrick (e.g., Chromosorb G AW), with formic acid-containing steam carrier gas 3. Porous glass beads (e.g., Corning GLC-100[e]), with formic acid-containing steam carrier gas
Polyhydric alcohols	Diatomaceous firebrick (e.g., Isolite LBK-28 modified with potassium fluoride or potassium hydrogen fluoride)
Amines	1. Sintered magnesia 2. Porous glass beads (e.g., Corning GLC-100), with pure steam or hydrazine-containing steam carrier gas 3. Deactivated diatomaceous firebrick (e.g., Chromosorb G AW), with hydrazine-containing steam carrier gas
Steroids	Porous glass beads (e.g., Corning GLC-100)

[a] Composition: 50% SiO_2, 45% Al_2O_3, 0.7% CaO; bulk density 0.8. Isolite Kôgyo Co., Osaka, Japan.

[b] Johns-Manville Products Corp., New York, N.Y.

[c] Waters Associates, Inc., Framingham, Mass.

[d] Shimazu Seisakujo, Kyoto, Japan.

[e] Corning Glass Works, Corning, N.Y.

E. Detector

The thermal conductivity detector (TCD) is unsuitable for SSC, which is
usually used for trace and rapid analysis. Also the TCD senses as noise the
pressure fluctuation of carrier gas, and, because of its detecting mechanism,
it cannot follow exactly the rapid elution signals of the samples.

The FID is very convenient in SSC. The FID used with steam carrier
gas shows little or no variation in sensitivity and in noise level as compared
with other ordinary carrier gases. To obtain better results in SSC the fuel
hydrogen is mixed with nitrogen gas before being introduced into the
detector nozzle. The flow rates of the hydrogen and the nitrogen are set
so as to maximize both the sensitivity and the signal-to-noise ratio depending on the carrier-steam flow rate. Even when formic acid or hydrazine
hydrate vapor is added to the carrier steam, the FID functions with normal
sensitivity and without a significant increase in noise level. The base-line
level, however, increases a little. This increase in base-line level may
be due to the fact that these corrosive vapors attack organic materials
(synthetic-resin tubing, silicone-rubber septum, etc.) used in the flow
system and carry the reaction products into the detector. The FID seems
to be almost entirely insensitive to the vapors of water, formic acid, and
hydrazine hydrate.

IV. PRACTICAL APPLICATIONS

A. General Remarks

The applications of SSC in analyzing various samples are shown in Figs. 2
through 22, which are typical chromatograms. In these examples steam
containing formic acid or hydrazine hydrate rather than pure steam was
used as the carrier gas. Most samples analyzed are injected in the form
of dilute aqueous solutions, or in the form of aqueous emulsions, or
aqueous suspensions in cases where the sample materials are not sufficiently
soluble in water. The smaller the sample size, the better the separation.
It is favorable, therefore, to use a sample size of less than 1 μl at concentrations of 0.1 to 0.01% in order to obtain a good separation, even in
the case of high-boiling or very polar samples. If the sample can be diluted
with water, the volume used can be precisely adjusted to the optimum for
the specific column and conditions used. The water used for diluting the
sample materials is not sensed by the FID, and the peak resolution may be
improved.

Even high-boiling or polar materials can be eluted very rapidly without
any marked tailing effect, as seen, for example, in Figs. 4, 5, 16, 19, and
22. In general, the eluted peaks of these samples are much higher than in

conventional GC. The increased heights of the elution peaks makes the detection limits much lower. For example, 0.1-ppm aqueous solutions of lower fatty acids and phenols, and 1-ppm solutions of higher fatty acids have been successfully analyzed by injecting 5 to 6 μl of these samples (Figs. 11 and 12).

Because the elution of the samples in SSC is very fast, the peak broadening of the sample that occurs during injection directly influences the eluted peak width, or the apparent HETP, and the resolution of the eluted peaks. In some cases, especially with highly polar samples, a single component yields two adjacent peaks. This is probably due to an irregular adsorption of the injected sample at the sample-evaporating port. Even in SSC sample injection must be managed with care.

Acidic columns are not always necessary even for analyzing strongly acidic samples if the steam carrier gas is mixed with formic acid vapor (10% w/w). Similarly most amine samples can be made to elute smoothly from nonbasic columns if the steam carrier gas contains hydrazine hydrate vapor (10 to 20% w/w). By using mixed carrier vapors organic acids and amines can be eluted from the same stationary solid, such as Chromosorb G AW (Johns-Manville).

Furthermore, in SSC, inorganic salts of organic acids can be analyzed directly not only with an acidic column treated with phosphoric acid and with a pure steam carrier gas but also with a neutral column (untreated with acids or bases) and a carrier gas containing formic acid vapor. Inorganic salts of amines can also injected directly if the carrier steam is mixed with hydrazine hydrate vapor.

When an adsorbent with an extremely low specific area, such as Corning GLC-100 (Corning Glass Works), is employed as the solid phase in SSC, the eluted peaks sometimes show anomalous profiles, which are probably due to the saturation of the adsorbent with sample molecules or a nonlinear adsorption isotherm (deviation from Henry's law). In such cases the peaks show the so-called saturation profiles, and the sample must be limited to a small amount —reduced, for example, to 10^{-8} g in the case of the Corning GLC-100 adsorbent in a 2-mm i.d. column. The Corning GLC-100 column must be 4 m or more in length in order to obtain satisfactory separation for mixed samples.

Some chromatograms shown in this chapter are inferior to those obtained with a well-designed conventional GC apparatus with respect to separation and resolution, even though most adsorbents used in these SSC systems were those that are readily available or can be easily modified. Adsorbents better suited to SSC will be found and employed in the near future, and more satisfactory chromatographic results may be anticipated.

B. Applications to a Variety of Samples

1. Low-Boiling Hydrocarbons

Although low-boiling hydrocarbons are among the most easily analyzable
samples with conventional GC systems, the diluted aqueous samples (as
emulsions or suspensions) can be analyzed more effectively with SSC
without such pretreatment as extraction. For these samples, even for
nongaseous hydrocarbons, SSC can be carried out at once with an ordinary
adsorbent — activated alumina. The retention volumes in this system are
fairly large for SSC, and the eluted peaks obtained are almost symmetrical.
Activated alumina (chromatographic use; specific area 400 m^2/g), used in
the SSC system with which the chromatograms of Figs. 2 and 3 were
obtained, seems to have too large a specific area and to adsorb too strongly.
When the activated alumina is deactivated slightly by calcining after adding
small amounts of certain salts (e.g., alkali-metal halides), the retention
times of hydrocarbons are short. It should be noted that the retention
volumes of low-boiling-point hydrocarbons in the GC system with an
activated-alumina column increase with increasing column temperature in
the range of 120 to 180°C and then begin to decrease as the column temperature exceeds 180°C (Fig. 23), whereas the dependence of the retention volume
on the column temperature is fortunately insignificant at a column temperature of about 180°C [33].

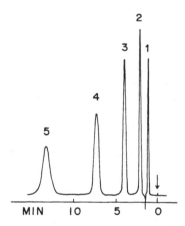

FIG. 2. Chromatogram of an aqueous emulsion of alkanes. Aluminum-
tube column, 3 m x 2 mm, packed with activated alumina, 40-60 mesh;
column temperature 122°C; steam carrier gas, 18 ml/min; FID detector;
sample size 1 μl of a 0.01% aqueous emulsion. Peaks: 1, n-pentane, (atten-
uated by one-half); 2, n-hexane; 3, n-heptane; 4, n-octane; 5, n-nonane.

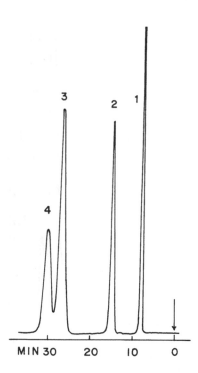

FIG. 3. Chromatogram of an aqueous emulsion of benzene derivatives. Aluminum-tube column, 3 m × 2 mm, packed with activated alumina, 40-60 mesh; column temperature 180°C; steam carrier gas, 17 ml/min; FID detector; sample size 1 μl of a 0.1% aqueous emulsion. Peaks: 1, benzene; 2, toluene; 3, p- and m-xylene; 4, o-xylene.

2. High-Boiling Hydrocarbons

There are certain advantages to the use of SSC in the accurate analysis of high-boiling hydrocarbons such as polynuclear hydrocarbons or polyphenyls. This is because SSC does not employ a liquid stationary phase unable to withstand the high temperatures necessary. Moreover, in SSC these high-boiling hydrocarbons are eluted rapidly even at a rather low column temperature; for example, p-quaterphenyl is eluted in about 3 min at 260°C from a column of Chromosorb P AW [18]. Typical chromatograms of a mixture of polyphenyls on a 2-m Chromosorb P AW 30/60 column and of a coal-tar fraction (300 to 400°C) on a 4-m HF-modified C-22 (Johns-Manville) column are shown in Figs. 4 and 5.

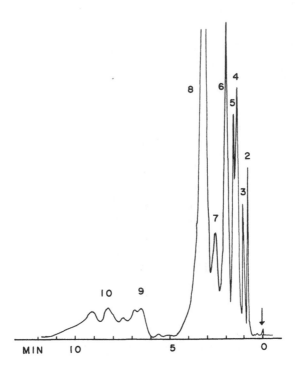

FIG. 4. Chromatogram of a coal-tar fraction (300 to 400°C). Glass-
tube column, 2 m x 3 mm, packed with C-22 modified with HF, noncoated,
80-100 mesh; column temperature 130 to 150°C; steam carrier gas, 19
ml/min; FID detector. Peaks: 2, naphthalene; 3, methylnaphthalene;
4, dimethylnaphthalene; 5, acenaphthene and trimethylnaphthalene; 6,
fluorene; 7, methylfluorene; 8, phenanthrene and anthracene; 9, methyl-
anthracene and methylphenanthrene; 10, fluoranthene.

     Samples of mixtures of polychlorobiphenyls (PCBs) having high boiling
points can also be analyzed with a column of Chromosorb P AW at a comparatively low column temperature and with short retention times, although
the separation is not perfect (see Fig. 6).

3. Monohydric Alcohols

Alcohols can be easily analyzed in most cases by GLC with a polar stationary phase. However, if the samples are in the form of dilute aqueous
solutions, which is often the case in natural products, quantitative and

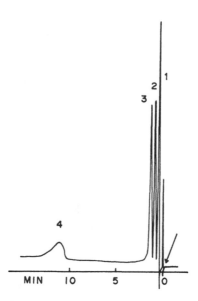

FIG. 5. Chromatogram of a 0.01% aqueous suspension of polyphenyls. Aluminum-tube column 2 m × 4 mm, packed with Chromosorb P AW, 30-60 mesh, noncoated; column temperature 225°C; steam carrier gas, 38 ml/min; FID detector; sample size 1 μl. Peaks: 1, o-terphenyl; 2, p-terphenyl; 3, triphenylene; 4, p-quaterphenyl.

qualitative GC analysis of alcohols is rather difficult because of the marked tailing of the water peak and the elution of the stationary liquid, which is decomposed by the water. By using SSC with powdered diatomaceous fire-brick (e.g., Isolite LBK-28) or porous silica (e.g., Porasil F) as an adsorbent it is possible to analyze alcohols as extremely dilute aqueous solutions (down to 0.3 ppm) by direct injection (see Fig. 7). Typical chromatograms obtained on diatomaceous firebrick and on porous silica columns are shown in Figs. 7, 8, and 9. At high temperatures the separation of these samples is not greatly reduced on a diatomaceous column, but on porous silica it is significantly reduced. Therefore for the rapid SSC analysis of alcohols it is favorable to employ such columns as diatomaceous firebrick at high temperatures. A column of diatomaceous firebrick powder packed in a borosilicate glass tube can be coiled by heating even after packing without any deterioration of column efficiency. The tailing effect with the alcohols in these SSC systems is quite insignificant, although all these samples have high polarity and strong adsorbability.

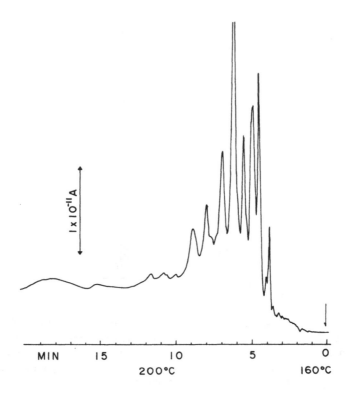

FIG. 6. Chromatogram of polychlorobiphenyls (Kaneclor-400). Glass-tube column, 4 m x 2 mm, packed with Chromosorb P AW, 30-60 mesh, noncoated; column temperature 160 to 200°C; steam carrier gas, 10 ml/min; FID detector.

4. Aldehydes, Ketones, Ethers, and Esters

Analysis of aldehydes, ketones, ethers, and esters is carried out with SSC columns similar to those used for monohydric alcohols. For these samples SSC gives good separation even with dilute aqueous solutions (see Fig. 10).

5. Glycols

By using diatomaceous firebrick powder modified with potassium fluoride or potassium hydrogen fluoride as the adsorbent, SSC can be used for the analysis of dilute aqueous solution of glycols. The elution of the glycol samples in the SSC system is very fast even at a comparatively low column temperature. For example, ethanediol and propanediol are separated in less than 2 min at a column temperature of 140°C.

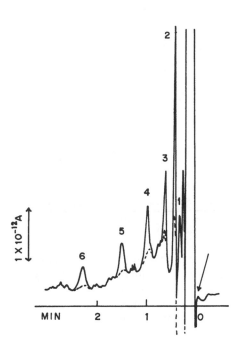

FIG. 7. Chromatogram of a 0.5-ppm aqueous solution of normal alcohols. Glass-tube column, 2.6 m × 2.2 mm, packed with firebrick (LBK-28) modified with HF, 30-60 mesh, noncoated; column temperature 130°C; steam carrier gas, 71 ml/min; FID detector; sample size 1 μl. Peaks: 1, methanol; 2, ethanol; 3, propanol; 4, butanol; 5, pentanol; 6, hexanol.

6. Carboxylic Acids and Phenols

It has been rather difficult to analyze carboxylic acids or phenols by GC without esterification because of their strong adsorbabilities, which cause marked tailing and long retention times (or no elution). This situation becomes much worse when GC is directly applied to aqueous samples of free carboxylic acids or phenols because of the interaction of the water with the stationary phase. However, SSC has showed its capability by analyzing not only free carboxylic acids or phenols but also their inorganic salts, even when the samples are aqueous solutions of less than 1 ppm.

Pure steam carrier gas and analytical columns of diatomaceous adsorbents (e.g., Isolite LBK-28 or Chromosorb P AW) modified with a small amount of phosphoric acid are used in SSC for this purpose. With formic acid in the carrier vapor columns of inactive adsorbents (e.g., Chromosorb GAW or Corning GLC-100) are useful for analyzing acidic samples.

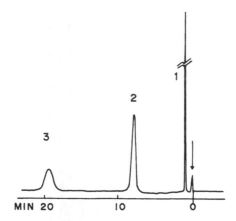

FIG. 8. Chromatogram of a 0.01% aqueous solution of a mixture of
(1) diacetone alcohol, (2) cyclohexanol, and (3) benzyl alcohol. Glass-tube
column, 4 m × 2.5 mm, packed with firebrick (LBK-28) modified with HF,
30-60 mesh, noncoated; column temperature 120°C; steam carrier gas,
57 ml/min; FID detector; sample size 1 μl.

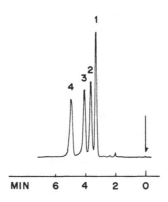

FIG. 9. Chromatogram of a 0.1% aqueous solution of butanols. Glass-
tube column, 2 m × 2 mm, packed Porasil F, 100-150 mesh, noncoated;
column temperature 120°C; steam carrier gas, 5 mg/min; FID detector;
sample size 1 μl. Peaks: 1, 2-methyl-2-propanol; 2, 2-butanol; 3,
isobutanol; 4, 1-butanol.

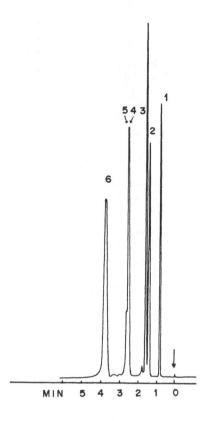

FIG. 10. Chromatogram of a 0.05% aqueous solution of a mixture of
(1) ethyl ether, (2) ethyl acetate, (3) methyl ethyl ketone, (4) methyl iso-
butyl ketone, (5) butyl acetate, and (6) 1-butanol. Glass-tube column,
2 m × 2 mm, packed with Porasil F, 100-150 mesh, noncoated; column
temperature 120°C; steam carrier gas, 9 ml/min; FID detector; sample
size 0.5 μl.

The acidic column used with pure steam carrier gas is prepared as
follows: The adsorbent is first packed into the column tube, and then a
2.5% aqueous solution of phosphoric acid is added to the column so that the
amount of acid is equal to about 3% of the adsorbent. The column is then
baked at 500°C for about 5 min. The column thus prepared is very stable
even when the column temperature is about 300°C. The amount of phosphoric
acid added to the column strongly influences the elution profile, the resolution, and the retention volume. Excessive tailing appears if the amount
of phosphoric acid is less than 3% of the adsorbent, whereas the retention
volume and the resolution are reduced appreciably if the amount of acid is

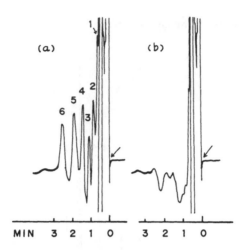

FIG. 11. Chromatograms of a 1-ppm aqueous solution of (a) lower fatty acids and (b) water used for diluting the samples by the same column. Glass-tube column, 4 m × 3.7 mm, packed with Chromosorb P AW baked after the addition of 3% H_3PO_4, 30-60 mesh, noncoated; column temperature 170°C; steam carrier gas, 81 ml/min; FID detector; sample size 5 μl. Peaks: 1, acetic acid; 2, propionic acid; 3, n-butylic acid; 4, n-valeric acid; 5, n-caproic acid; 6, enanthic acid.

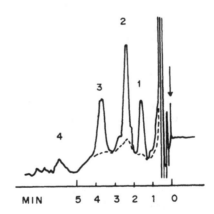

FIG. 12. Chromatogram of a 1-ppm (acid content) aqueous solution of potassium salts of higher fatty acids. Glass-tube column, 4 m × 3.7 mm, packed with Chromosorb P AW baked after the addition of 3% H_3PO_4, 30-60 mesh, noncoated; column temperature 230°C; steam carrier gas, 110 ml/min; FID detector; sample size 6 μl. Peaks: 1, potassium laurate; 2, potassium myristate; 3, potassium palmitate; 4, potassium stearate; broken curve, water only.

244

more than 3% of the adsorbent. In SSC with phosphoric acid-modified
columns, C_2 to C_{18} fatty acids, C_2 to C_7 dicarboxylic acids, benzoic
acid, salicylic acid, phenol, cresol, and alkylphenols — as well as their
inorganic salts — can be analyzed directly by injecting the samples in the
form of 1000- to 0.1-ppm aqueous solutions as well as in solutions of
ordinary organic solvents. Typical chromatograms of these samples are
shown in Figs. 11, 12, and 13.

Figures 14 through 17 show some examples in which acidic carrier gas
was used with columns of unmodified Chromosorb G AW and Corning GLC-
100. With a Chromosorb G AW column and a carrier steam containing 10%
(w/w) formic acid, both lower fatty acids and higher fatty acids are eluted
in an order related strictly to their boiling points. In a GC system using
Corning GLC-100 as the stationary solid and a carrier steam containing
the same concentration of formic acid, lower fatty acids (acetic acid to
enanthic acid) are eluted in reverse order with relation to their boiling
points (Fig. 15), whereas higher fatty acids (lauric acid to stearic acid)
are eluted in the regular order. The profiles of the eluted peaks in the SSC
system in which acidic steam carrier gas is used are, in general, fronting.
This effect is particularly significant in the case of Corning GLC-100 at a
rather low column temperature. This fronting effect may be due to the
nonlinear isotherms for adsorption of the sample components on the
stationary solids [34, 35], that is, due to type III adsorption isotherms,
which are convex in the pressure axis, according to the classification of
Brunauer et al. [36]. Because of these nonlinear isotherms the profile

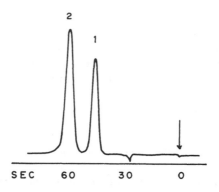

FIG. 13. Chromatogram of a 0.01% aqueous solution of cresols.
Glass-tube column, 3 m × 3.7 mm, packed with Chromosorb P AW baked
after the addition of 3% H_3PO_4, 30-60 mesh, noncoated; column temperature
144°C; steam carrier gas, 60 ml/min; FID detector; sample size 0.3 μl.
Peaks: 1, o-cresol; 2, m- and p-cresol.

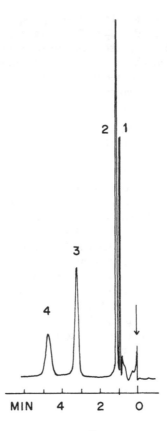

FIG. 14. Chromatogram of a 0.05% aqueous solution of lower fatty
acids obtained with an acidic carrier steam. Glass-tube column, 2 m x 2
mm, packed with Chromosorb G AW, 30-60 mesh, noncoated; column
temperature 125°C; steam containing 10% formic acid, 96 ml/min; FID
detector; sample size 1 μl. Peaks: 1, acetic acid; 2, propionic acid;
3, n-caproic acid; 4, enanthic acid.

of the eluted peak becomes wider and the position of the top of the peak
shifts backward when too big a sample is injected [2]. In this case,
therefore, to measure retention volumes exactly, the sample size must be
limited to some value, the magnitude of which depends on the sample, the
adsorbent, the column temperature, and the amount of formic acid in the
steam carrier gas; for example, a stearic acid sample is limited to about
10^{-8} when a 2-mm i.d. column of Corning GLC-100 is used with a steam
carrier gas containing 10% formic acid at a column temperature of 150°C,
whereas with Chromosorb G AW it is possible to introduce a larger sample,

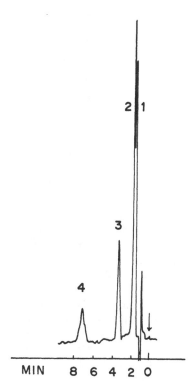

FIG. 15. Chromatogram of an aqueous solution of lower fatty acids
obtained with an acidic carrier steam. Glass-tube column, 4 m × 2 mm,
packed with Corning GLC-100, 60-80 mesh, noncoated; column temperature
130°C; steam containing 10% formic acid, 10 mg/min; FID detector;
sample size 0.5 μl. Peaks: 1, enanthic acid; 2, n-caproic acid; 3,
propionic acid; 4, acetic acid.

that is, about 10^{-7} at the same temperature and carrier flow. In general,
the higher the molecular weight of the sample and the higher the column
temperature, the larger the sample may be.

The preferred content of formic acid in the steam carrier gas is about
10% w/w. When there is less than 5% of formic acid in the steam, the
peak profiles of higher fatty acids on the Corning GLC-100 adsorbent are
appreciably widened and show significant fronting. On the other hand, with
Chromosorb G AW, less formic acid in the carrier steam increases the
retention time and causes a marked tailing in the elution profiles for higher
fatty acids. A steam carrier gas containing more than 10% formic acid
gives no increase in the chromatographic efficiency for either Corning
GLC-100 or Chromosorb G AW, but it has some disadvantages owing to the
corrosive action of the acid on the apparatus at high temperatures.

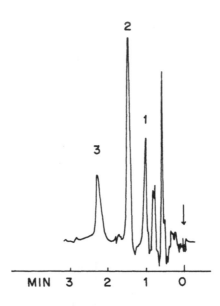

FIG. 16. Chromatogram of higher fatty acids obtained with a formic acid-containing carrier steam. Glass-tube column, 4 m × 2 mm, packed with Corning GLC-100, 60-80 mesh, noncoated; column temperature 200°C; steam containing 10% formic acid, 10 mg/min; FID detector; Peaks: 1, myristic acid; 2, palmitic acid; 3, stearic acid.

In SSC with formic acid added to the carrier steam and inactive-adsorbent columns, dicarboxylic acids cannot be eluted. They can be eluted with a pure steam carrier gas on acidic columns modified with phosphoric acid [18].

7. Amines

For the SSC analysis of various amine samples inactive adsorbents (e.g., porous glass, sintered magnesia, or inactivated diatomaceous firebrick) are used in combination with pure steam or steam containing 10 to 20% (w/w) hydrazine hydrate. When pure steam is used as a carrier gas for lower alkylamines, fairly marked tailing is present in the elution peaks even with columns of very inactive or slightly basic adsorbents, such as Corning GLC-100 or sintered magnesia modified with potassium hydroxide. On the other hand, with steam containing hydrazine hydrate the lower alkylamines are eluted from a column of unmodified Chromosorb G AW. Figures 18 and 19 show typical chromatograms obtained with an SSC system using steam carrier gas containing hydrazine hydrate. The eluted peaks of lower monoalkylamines up to

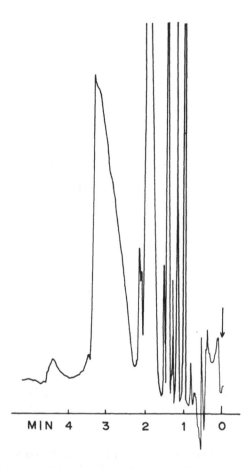

FIG. 17. Chromatogram of an aqueous solution of soap. Chromatographic conditions are the same as in Fig. 16.

n-octylamine have an almost symmetrical profile, and the resolution is
satisfactory. Mixed samples of monoalkylamines lower than propylamine
are not completely separated from each other. Di- and trialkylamines,
such as di- and trimethylamine and di- and tributylamine, are also separated
from each other or from the corresponding monoalkylamine in this SSC
system. Higher fatty amines, such as laurylamine, cetylamine, and stearylamine, can be introduced as samples in this SSC system with a Chromosorb
G AW column and hydrazine-containing steam carrier gas, although the
eluted peaks are slightly distorted by a tailing effect that appears to increase
with an increase in the molecular weight of the amines and to decrease with
an increase in the column temperature. Even for these higher fatty amines

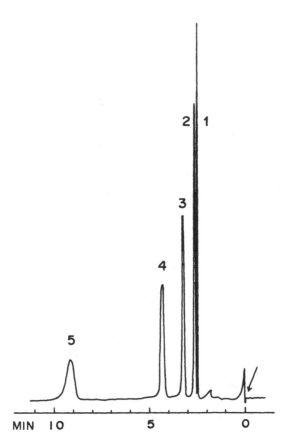

FIG. 18. Chromatogram of a 0.1% aqueous solution of lower alkylamines.
Glass-tube column, 4 m x 2 mm, packed with Chromosorb G AW, 30-60
mesh, noncoated; column temperature 130°C; steam containing 20%
hydrazine hydrate, 9 ml/min; FID detector; sample size 0.5 μl. Peaks:
1, n-propylamine; 2, n-butylamine; 3, n-amylamine; 4, n-hexylamine;
5, n-octylamine.

the retention times are rather short, as shown in Fig. 19. Aromatic
amines, such as pyridine, aniline, toluidine, and quinoline, also are
satisfactorily eluted under conditions similar to those used for fatty amines.
The hydrazine-containing steam carrier gas can cause the inorganic salts
of these amines to elute at the same retention times as those of the corresponding free amines.

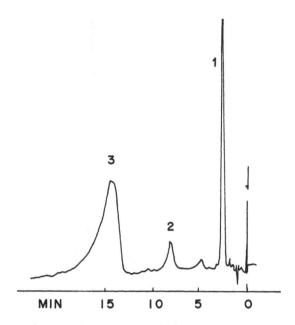

FIG. 19. Chromatogram of higher fatty amines. Glass-tube column,
4 m x 2 mm, packed with Chromosorb G AW, 30-60 mesh, noncoated;
column temperature 200°C; steam containing 10% hydrazine hydrate,
10 mg/min; FID detector; Peaks: 1, laurylamine; 2, cetylamine; 3,
stearylamine.

When a Corning GLC-100 column is used with a steam carrier gas
containing 10 to 20% hydrazine hydrate, the elution of higher fatty mono-
amines is so rapid that the eluted peaks cannot be separated from each
other, although this SSC system is suitable for the analysis of more polar
or more basic materials, such as aminoalcohols and diamines. Figure 20
shows a typical chromatogram of a mixture of ethylenediamine and ethanolamine. On the other hand, it seems to be difficult to elute these materials
on a Chromosorb G AW column even in combination with such a basic
carrier gas.

The preferred content of hydrazine hydrate in the steam carrier gas is
10 to 20% (w/w). Hydrazine hydrate is so corrosive that it may attack any
part of the chromatograph. Increasing the hydrazine hydrate content above
20% does not improve resolution. When a steam carrier gas containing less
than 10% hydrazine hydrate is used, the elution of amine samples becomes
rather slow and the profiles of the eluted peaks are skewed.

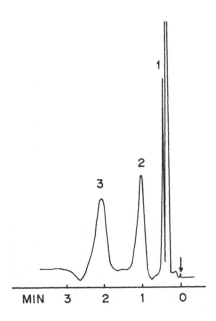

FIG. 20. Chromatogram of a 0.1% aqueous solution of a mixture of
(1) butylamine, (2) ethylenediamine, and (3) ethanolamine. Glass-tube
column, 4 m × 2 mm, packed with Corning GLC-100, 60-80 mesh, noncoated; steam containing 10% hydrazine hydrate, 19 ml/min; FID detector;
sample size 1 μl.

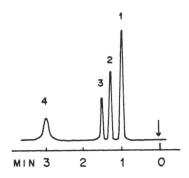

FIG. 21. Chromatogram of a 0.1% aqueous solution of aromatic amines
obtained with an ammonia-containing carrier steam. Glass-tube column,
4 m × 2 mm, packed with Chromosorb G AW, 30-60 mesh, noncoated;
column temperature 110°C; steam containing 5% (w/w) ammonia, 25 ml/min;
FID detector; sample size 0.2 μl. Peaks: 1, pyridine; 2, aniline; 3,
p-toluidine; 4, quinoline.

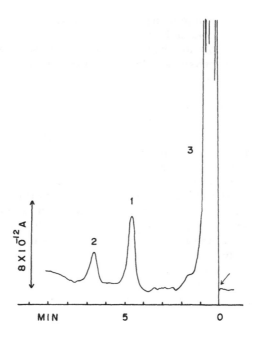

FIG. 22. Chromatogram of an aqueous sample of cholesterol and stigmasterol. Glass-tube column, 4 m x 2 mm, packed with Corning GLC-100, 60-80 mesh, noncoated; column temperature 200°C; carrier gas, 22 ml/min; FID detector; sample size 1 μl. Peaks: 1, cholesterol (saturated aqueous solution); 2, stigmasterol ($\sim$10^{-5}-g/ml aqueous suspension); 3, solvent.

Ammonia can also be added to the steam carrier gas to facilitate the analysis of amine samples. Ammonia cannot be expected to be too useful because of its comparatively weak adsorbability. An ammonia-containing carrier steam is prepared by using a 5 to 10% aqueous solution of ammonia in the pumping material for the carrier vapor. Figure 21 shows a typical chromatogram of aromatic amines obtained with an ammonia-containing carrier steam.

8. Steroids

An extremely inactive adsorbent may be preferred in the SSC analysis of steroid samples. Figure 22 shows a typical chromatogram of a dilute aqueous sample of a mixture of cholesterol (as a saturated aqueous solution) and stigmasterol (as a 10^{-5}-g/ml aqueous suspension). This chromatogram

was obtained in an SSC system using a Corning GLC-100 column and a pure
steam carrier gas, at a column temperature of 200°C. The SSC technique
seems to be rather suitable for the analysis of steroid samples as dilute
aqueous solutions or suspensions. However, a better solid phase is needed.

V. CHROMATOGRAPHIC MECHANISM

In contrast to common experience in GC, the SSC analysis of low-boiling
hydrocarbons on an activated-alumina column presents an anomalous
relation between retention volume and column temperature: the retention
volume increases with increasing column temperature up to 180°C and then
begins to decrease. At a column temperature of less than about 120°C the
retention volume decreases so abruptly that separation becomes impossible.
Figure 23 shows semilogarithmic plots of retention volumes for typical
samples as a function of the reciprocal of the absolute temperature of the
column. This anomaly in the relation between retention volume and temperature can be reasonably explained in terms of an interaction between the
sample molecules and the adsorbent through a water layer on the surface of
the adsorbent. This water layer is formed by multilayer physical adsorption
at comparatively high pressures [26]. When the column temperature is

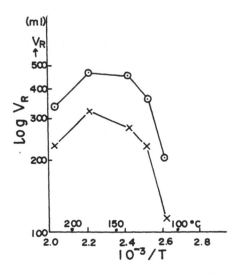

FIG. 23. Temperature dependence of retention volume in SSC of
hydrocarbons. Activated-alumina column. Samples: o-xylene (circles)
and n-nonane (crosses).

raised, the thickness of the water layer on the adsorbent surface decreases. This results, in turn, in an increase in the heat of adsorption of the sample, if the sample molecules are assumed to be adsorbed on the water layer or in the water layer under the potential field of an attractive force exerted by the true surface of the adsorbent. It is understood, therefore, that in the well-known formula showing the relation between retention volume, heat of adsorption, and column temperature, $V_R = C \exp(\Delta H_a/RT)$, the retention volume V_R of the sample increases with the column temperature T if the increase in the heat of adsorption ΔH_a of the sample exceeds the increase in RT.

In ordinary GSC, where the heat of adsorption ΔH_a is assumed to be a constant, depending only on the surface nature of the adsorbent and the nature of the sample molecule, but independent of column temperature, the relation between the logarithmic values of retention volumes and the reciprocal of the absolute temperatures of the column must be strictly linear. In SSC, however, the linearity of the relation may not always be expected because the heat of adsorption is no longer assumed to be a constant. Even in the case where the retention volume decreases monotonically with increasing column temperature, as in an SSC system of a diatomaceous adsorbent and lower alcohol samples, the heats of adsorption of the samples cannot be estimated from the slope of the curve showing the relation log V_R versus $1/T$. In an extremely anomalous case, such as an activated-alumina column and low-boiling-point hydrocarbon samples, the slope becomes negative in some column-temperature intervals and the adsorption heat may be unreasonably assumed to be positive.

Multilayer physical adsorption, which is considered to be a sort of surface condensation, has been expected theoretically and was shown experimentally to exist. However, most experiments for obtaining adsorption isotherms showing the existence of multilayer physical adsorption have been carried out only at room temperature. Only a few experimental results at high temperatures and high pressures of adsorbates have been reported. Figure 24 shows the adsorption isotherms of water vapor obtained by the author with a diatomaceous adsorbent (Chromosorb P AW) and water vapor at a temperature range of 125 to 145°C and a reduced pressure (P/P_0) range of 0 to 0.8 (P is the pressure of steam and P_0 is the saturation pressure of steam at the temperature). The Y-axis in Fig. 24 shows the surface coverage, which is equal to the number of layers of adsorbate molecules adsorbed to the surface. The surface coverage in this figure involves chemisorbed water layers. Since under the usual SSC conditions the reduced pressure P/P_0 is about 0.3 to 0.7, the surface of the adsorbent is covered, in this case, with four to six layers of chemisorbed water. It also becomes clear that the isosteric differential heat of adsorption for the water molecules at the top of the adsorbed layer is similar in value to the heat of liquefaction of water, although the isosteric heats of adsorption are usually estimated

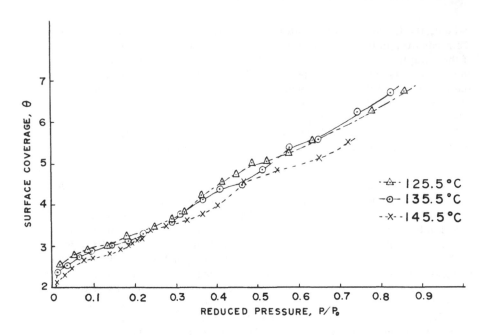

FIG. 24. Adsorption isotherms for water vapor on Chromosorb P AW.

from the adsorption isotherms obtained experimentally. It is understood, therefore, that for the molecules at the top of the adsorbed layer the interactive influence from the true surface of the adsorbent becomes much smaller and the principal part of the heat of adsorption comes from the energy of interaction with the surface layer of the adsorbed water. Taking into account only the interaction between the surface layer of water and adsorbed sample molecules, and neglecting the interaction between the true surface of the adsorbent and the adsorbed sample molecules in the case of SSC the heats of adsorption of the adsorbed sample molecules may be nearly equal to or lower than their heats of liquefaction if the molecules are almost insoluble in water. It has been reported that the heat of adsorption of such molecules is rather lower than the heat of liquefaction [37, 38]. Even if the adsorbed sample materials are soluble in water, the heat of adsorption of the sample on the adsorbed water layer is assumed to be, at most, the sum of the heat of liquefaction and the heat of solubilization. On the other hand, for vaporizable organic materials, the heat of adsorption onto the clean surfaces of ordinary inorganic adsorbents is reported to be several tens of kilocalories per mole, and such adsorptions are considered to be chemisorptive [7, 8]. Therefore it is supposed that the presence of a multilayer of water on the surface makes the adsorption energies of foreign adsorbed molecules significantly smaller and the water

layer on the SSC adsorbent, in turn, can make the elution of most organic
sample materials very fast. Of course, SSC adsorbents must be chosen to
suit the sample materials: that is, considering whether the samples are
nonpolar, highly polar, low-boiling, high-boiling, acidic, or basic, and
so on. This is because the interactive influence on the adsorbed sample
molecules from the real surface of the adsorbent must persist to some
extent after the adsorbed water layer is formed on the surface.

It is no wonder that a very polar and rather active carrier gas such
as water is more effective than a nonpolar vapor for analyzing polar as well
as nonpolar samples. Polar sites on adsorbent preferentially adsorb the
polar carrier vapor, which covers the polar sites so completely that the
adsorbent surface is fully deactivated for the polar samples. The significant
ability of formic acid mixed with the steam carrier gas to rapidly elute
organic acids may be due to the fact that formic acid is adsorbed preferentially to the basic sites (more basic than formic acid) on the adsorbent.
The case of hydrazine hydrate and the acidic sites is exactly analogous.
Under such favorable conditions, when the basic or acidic sites are fully
neutralized, the sample molecules are hindered to a great extent from
being absorbed on the basic or acidic sites on the real surface. However,
when the acidity or basicity of the sample is greater than the acidity or
basicity of the carrier material, an exchange of adsorption may be brought
about between the sample and the adsorbed carrier molecules on the basic
or acidic sites of the adsorbent. Under such conditions the sample molecules
cannot be easily eluted, as seen in the SSC system where dibasic carboxylic
acids (e.g., malonic or succinic acid) are analyzed with a non-acid-modified
column and a steam carrier gas containing formic acid, which is less
acidic than the sample materials. The terms "basicity" and "acidity" used
in this explanation are to be taken in the most extensive sense.

When an active carrier vapor is used in the mobile SSC phase, a particular effect in the mobile phase, in addition to the effect in the stationary
phase mentioned above, may be expected. If the carrier-vapor molecules
interact with the sample molecules present in gas phase and the heat of
interaction (or heat of reaction) is negative, the apparent heat of adsorption
of the sample molecules to the stationary surface must be reduced by an
amount corresponding to the heat of interaction in the gas phase. This
intermolecular interaction in the gas phase is considered to be fairly strong
when both carrier and sample molecules are of high polarity, or of high
polarizability, or have polar groups. It is well known that such molecules
as water, ammonia, and amines have high second virial coefficients (or
mutual virial coefficients in the case of mixtures) near the liquefaction point
[39-41]. In the SSC experiments there is as yet no straightforward evidence
of the interactive effect between sample and carrier molecules in the gas
phase.

When the gas phase is at an extremely high pressure, the intermolecular interaction in the gas phase may be so significant that the sample molecules can be considered to dissolve in the gas phase, even if both the sample and carrier gas molecules are inactive and nonpolar, as seen in high-pressure gas chromatography [42, 43]. It may be said that SSC and high-pressure gas chromatography are similar to liquid chromatography in that all three involve strong intermolecular interactions between sample and carrier molecules.

VI. CONCLUSIONS

Steam carrier gas-solid chromatography promises several advantages over conventional techniques of gas chromatography. It is a new technique, and further development work is needed to bring it to the level of the older techniques. Among the areas that invite further work is the use of steam carrier gas at higher temperatures and pressures. Furthermore, the direct combination of liquid chromatography with SSC, or vapor chromatography in a more general sense, will be possible since the SSC system can employ a liquid pump. The adsorbents presently used in SSC have given some excellent results, but there is much room for improvement, especially in finding new adsorbents and pretreatments, and tailoring them to specific analytical requirements.

Among the most attractive features of SSC is its ability to accept samples in dilute aqueous solution or suspension and also as salts without sample preparation. The technique should have useful applications in biochemistry, medical science, pharmacology, the food industry, and especially in the analysis of environmental pollution.

REFERENCES

1. J. C. Giddings, Anal. Chem., 36, 1170 (1964).

2. A. V. Kiselev and Y. I. Yashin, Gas-Adsorption Chromatography, Plenum Press, New York-London, 1969.

3. A. V. Kiselev, in Gas Chromatography 1964, A. Goldup, ed., Institute of Petroleum, London, 1965, p. 238.

4. O. L. Hollis, Anal. Chem., 38, 309 (1966).

5. C. Vidal-Madjar and G. Guiochon, Nature, 215, 1372 (1967).

6. V. Mahadevan and L. Stenroos, Anal. Chem., 39, 1952 (1967).

7. W. H. Wade, S. Teranishi, and T. L. Durham, J. Phys. Chem., 69, 590 (1965).

8. Y.-F. Yu Yao, J. Phys. Chem., 69, 3930 (1965).

9. A. Nonaka, Bunseki Kagaku, 16, 260 (1967).

10. A. Nonaka, Bunseki Kagaku, 16, 1166 (1967).

11. A. Nonaka, Bunseki Kagaku, 17, 91 (1968).

12. A. Nonaka, Bunseki Kagaku, 17, 944 (1968).

13. A. Nonaka, Bunseki Kagaku, 17, 1215 (1968).

14. T. Morii and H. Arimoto, Bunseki Kagaku, 18, 900 (1969).

15. A. Nonaka, Sekiyu Gakkai-shi, 14, 86 (1971).

16. A. Nonaka, Bunseki Kagaku, 20, 416 (1971).

17. A. Nonaka, Bunseki Kagaku, 20, 422 (1971).

18. A. Nonaka, Anal. Chem., 44, 271 (1972).

19. A. Nonaka, Anal. Chem., 45, 483 (1973).

20. A. Nonaka, Bunseki Ki-ki, 10, 236 (1973).

21. I. Halász and E. Heine, in Advances in Chromatography, J. C. Giddings and R. A. Keller, eds., Dekker, New York, 1967, p. 239.

22. H. S. Knight, Anal. Chem., 30, 2030 (1958).

23. A. Davis, A. Roaldi, and L. E. Tufts, J. Gas Chromatogr., 2, 306 (1964).

24. L. H. Phifer and H. K. Plummer, Jr., Anal. Chem., 38, 1652 (1966).

25. B. L. Karger and A. Hartkopf, Anal. Chem., 40, 215 (1968).

26. D. M. Young and D. Crowell, Physical Adsorption of Gases, Butterworths, London, 1962.

27. T. Tsuda, N. Tokoro, and D. Ishii, J. Chromatogr., 46, 241 (1970).

28. J. Van Rysselberge and M. Van Der Stricht, Nature, 193, 1281 (1962).

29. R. L. Pecsok and E. M. Vary, Anal. Chem., 39, 289 (1967).

30. J. C. Giddings, J. Chromatogr., 3, 443 (1960).

31. J. C. Giddings, Dynamics of Chromatography, Part 1, Dekker, New York, 1965.

32. J. J. Kirkland, Anal. Chem., 43, 37A (1971).

33. P. F. McCrea and J. H. Purnell, Anal. Chem., 41, 1922 (1969).

34. D. de Vault, J. Am. Chem. Soc., 65, 532 (1943).

35. L. D. Belyakova, A. V. Kiselev, and N. V. Kovaleva, Bull. Soc. Chim. Fr. , 285 (1967).

36. S. Brunauer, The Adsorption of Gases and Vapours, Vol. 1, Princeton University Press, Princeton, N.J. , 1945.

37. B. L. Karger, P. A. Sewell, R. C. Castells, and A. Hartkopf, J. Colloid Interface Sci. , 35, 328 (1971).

38. B. L. Karger, R. C. Castells, P. A. Sewell, and A. Hartkopf, J. Phys. Chem. , 75, 3870 (1971).

39. J. D. Lambert, G. A. H. Roberts, J. S. Rowlinson, and V. J. Wilkinson, Proc. Roy. Soc. (London), A196, 113 (1949).

40. J. H. P. Fox and J. D. Lambert, Proc. Roy. Soc. (London), A210, 557 (1952).

41. G. S. Kell, G. E. Mclaurin, and E. Whalley, J. Chem. Phys. , 48, 3805 (1968).

42. S. T. Sie and G. W. A. Rijnders, Separation Sci. , 2, 755 (1967).

43. J. C. Giddings, M. N. Myers, and J. W. King, J. Chromatogr. Sci. , 7, 276 (1969).

AUTHOR INDEX

Numbers in parentheses are reference numbers and indicate that an author's work is referred to although his name is not cited in the text. Underlined numbers give the page on which the complete reference is listed.

SUBJECT INDEX

A

α, see Separation factor

Acetone, diffusion
  in Ar, 135
  in He, 130
  in H_2, 133

Acids, see Carboxylic acids

Affinity chromatography, 69, 72-74, 78-79
  packings for, 74

Aflatoxins, 2-3

Alcohols, analysis, by gas solid chromatography, 238-240

Aldehydes, analysis, by gas solid chromatography, 240

Amines, analysis, by gas solid chromatography, 248-253

Ammonia, diffusion, in He, 125

Argon, diffusion in, 134-135, 138
  diffusion in He, 124-125, 138

Aroclors, see Polychlorinated biphenyls and Pollution analysis

B

Base-line studies, 19-21

Beer's law, 18
  callibration plot, 21

Benzene, diffusion
  in Ar, 134
  in CO_2, 136
  in H_2, 120-121

(Benzene, diffusion)
  in He, 127-128
  in N_2, 133

Binary diffusion, see Gaseous diffusion

Biphenyls, see Polychlorinated biphenyls

Bromobenzene, diffusion, in He, 131

1-Bromobutane, diffusion, in He, 130

2-Bromobutane, diffusion, in He, 130

2-Bromo-1-chloropropane, diffusion, in He, 131

Bromoethane, diffusion, in He, 130

1-Bromohexane, diffusion, in He, 130

2-Bromohexane, diffusion, in He, 130

3-Bromohexane, diffusion, in He, 131

1-Bromopropane, diffusion, in He, 131

2-Bromopropane, diffusion, in He, 131

Butane, diffusion
  in Ar, 134
  in H_2, 120
  in He, 127, 138
  in N_2, 132

1-Butanol, diffusion
  in H_2, 121
  in He, 129

2-Butanol, diffusion, in H_2, 121

271

G

Gas chromatography, <u>see also</u>
    Gaseous diffusion and Gas-
    solid chromatography
  electrometers for, 177-222
  organic pollutants, nonpesticide,
    141-168
Gaseous diffusion, 99-138
  apparatus for measurement, 105-
    106
  arrested-ellution method, 111
  binary coefficient, 120-138
  in capillary columns, 109
  closed-tube technique, 106
  collision cross section, 116
  column configuration, effects on
    measurements, 114, 118
  critical volume, dependence on,
    119
  dead volume, effect on measure-
    ment, 114
  dense gas, 138
  error
    in estimation, 111-112
    in measurement, 105-106
  estimation, from molar volume
    and molecular weight, 118
  evaporation-tube technique, 106
  frontal analysis, 108
  Golay equation, use of, 102-103
  high pressure, 119
  isomers, 119
  literature review, 107-119
  mass balance in, 100-101
  from mass transfer data, 110
  optimum measurement, 112
  packed column, 107, 112
  precision, 119
  pressure dependence, 112, 117-
    118
  stopped-flow measurement, 110
  temperature dependence, 109-110,
    113-114, 118

(Gaseous diffusion)
  theory, 100-103
  two-bulb apparatus, 106-107
  velocity, optimum, 102
Gas-solid chromatography, 223-258
  adsorbants, 230-233
  adsorption, isotherm, 256
  applications, 234-254
  column configuration, 232
  detectors for, 234
  features, 225-227
  heat of adsorption, 255
  injection port, 230
  liquid pump, 228-229
  mechanism, 254-258
  multilayer adsorption, 255-257
  steam boiler, 227-228
  steam generators, 227-229
  temperature
    control, 232
    dependence, 254-255
GC, <u>see</u> Gas chromatography
Gel-permeation chromatography,
    35-48, 69, 80
  applications, 36-37
  calibration standards, 46-48
  cellulose, 37-44
    derivatives, 44-49
  column packing, 37-38
  efficiency in, 67
  limitations, 49-51
  methodology, 35-36
  molecular weight determination,
    51-55
  solvents, 37-38
Giddings coupled equation, 109
GLC, <u>see</u> Gas chromatography
Glycols, analysis by gas-solid
    chromatography, 240
Golay equation, 102-103
GPC, <u>see</u> Gel-permeation chromato-
    graphy
Gradiant elution, 83
GSC, <u>see</u> Gas-solid chromatography